Environmental Biotechnology

Environmental Biotechnology

Theory and Application

Second Edition

by

Gareth M. Evans
Judith C. Furlong

Taeus Biotech, Banff, Banffshire, UK

WILEY-BLACKWELL

A John Wiley & Sons, Ltd., Publication

This edition first published 2011
© 2011 John Wiley & Sons Ltd

Wiley-Blackwell is an imprint of John Wiley & Sons, formed by the merger of Wiley's global Scientific, Technical and Medical business with Blackwell Publishing.

Registered office: John Wiley & Sons Ltd, The Atrium, Southern Gate, Chichester, West Sussex, PO19 8SQ, UK

Editorial Offices:
9600 Garsington Road, Oxford, OX4 2DQ, UK
The Atrium, Southern Gate, Chichester, West Sussex, PO19 8SQ, UK
111 River Street, Hoboken, NJ 07030-5774, USA

For details of our global editorial offices, for customer services and for information about how to apply for permission to reuse the copyright material in this book please see our website at www.wiley.com/wiley-blackwell

Library of Congress Cataloging-in-Publication Data

Evans, Gareth (Gareth M.)
    Environmental biotechnology : theory and application / by Gareth M. Evans, Judith C. Furlong.
        p. cm.
    Includes bibliographical references and index.
    ISBN 978-0-470-68418-4 (cloth) – ISBN 978-0-470-68417-7 (pbk.)
    1. Bioremediation. I. Furlong, Judith C. II. Title.
    TD192.5.E97 2011
    628.5–dc22

                                                                        2010032675

A catalogue record for this book is available from the British Library.

This book is published in the following electronic formats: ePDF: 978-0-470-97514-5; Wiley Online Library: 978-0-470-97515-2; ePub: 978-0-470-97538-1

Typeset in 10/12pt Times-Roman by Laserwords Private Limited, Chennai, India
Printed in Singapore by Ho Printing Singapore Pte Ltd

First Impression   2011

This book is dedicated with much love to our respective parents, partners and pooches.

Contents

Foreword

Environmental biotechnology has come of age and many of the technologies have
developed at a surprising rate both scientifically and perhaps more importantly as
practical techniques too. On both counts then, it's appropriate that there is now
to be a second edition of this book. Industry has been struggling, particularly
over recent years, with the need for increasing compliance with new laws and
a changed public perspective and all of this has made 'green' technologies very
attractive. But this can only work in real terms when those technologies are cost
effective and the emphasis is on both 'cost' and 'effective'. It is something that,
happily, both sides, industry and technology providers, seem increasingly to have
woken up to.

Carbon reduction could be a massive force for change. In the water industry
alone sludge thickening is a hideously energy-intensive process and accounts
for something like a third of the capital costs at a plant and half of its operating
costs. If you once start looking at ways to reduce carbon consumption then you're
bound to be finding ways to reduce those costs too and so initiatives such as the
UK's recent Climate Change Act and the Carbon Reduction Commitment have
major implications and are going to drive a major step-change across industry as
a whole. It could also be a massive spring-board for sustainable growth too.

Ecosystems are unbelievably complex, but the basic principles that drive them
are equally simple and it is precisely this area that is the natural realm of
environmental biotechnology and why the second edition of this book is so
timely. It provides an excellent framework for understanding the fundamentals
and biochemical processes that underpin the practical, and a cogent exposition
of those practical applications themselves. It is a complete introduction to the
broad church, that is modern environmental biotechnology and as such will be
of great value to undergraduates of course, studying the subject, but also to other
professionals in the wide and growing range of industries to which the subject
is becoming increasingly relevant.

With its logical structure, clear and authoritative explanations and fundamentally readable style, I whole-heartedly recommend it to any students, researchers
and environmental managers, in fact anyone who is looking to understand this
important science and how it will, I hope, play a big part in shaping our low
carbon future.

Dr Dene Clackmann
Principal of Clackmann Associates and Chair of the Carbon Commune Group

Preface

When we began work on the first edition of this book back in 2001, we set out to present a fair reflection of the practical biological approaches that were then currently being employed to address environmental problems, and to provide the reader with a working knowledge of the science which underpinned them. It was a straightforward goal and one which, like the book itself, sprang out of our Environmental Biotechnology modules at the University of Durham, but as we said at the time, this was never intended to be just another 'book of the course'. That thinking remains in this version, but we have been given the rare opportunity in it of being able to revise and update the content to once again, we hope, give a fair and honest overview of the real world applications of this fascinating branch of environmental management, ten years on.

It has been an interesting journey, particularly in respect of the outcomes of technologies and techniques that were innovative and new then – seeing which of our predictions came true and those for which we were wide of the mark. On balance, we feel the former satisfyingly outnumber the latter, but modesty prevents us from banging that particular drum too loudly; more objective minds than ours should make that call, if they so wish.

We received many useful comments from the many reviews and reviewers of the first edition, which was it seems, gratifyingly well received and a good number of their suggestions have helped to shape the changes that we have made. In the process of writing this update, we bore two things in particular, in mind. Firstly, from the comments we have had from successive waves of our own students and some of those who had read the earlier version, we believed that the fundamental approach we have adopted to the subject, works. That in itself is less of a unique insight than the result of the happy accident of our respective backgrounds, which so perfectly mix the academic and the practical, making 'theory and application' a natural focus as much as an appropriate title. Secondly, we were reminded of the words of an editor of our acquaintance: the most powerful drive known to our species is not for survival, nor to procreate, but to alter someone else's copy. We decided that unless it was really justified, we were not going to change our own.

Consequently, although all the case studies are new – as befits the progress that has been made in this field – the familiar shape of the original remains.

We have retained the logical structure we adopted at the outset addressing technologies in as cohesive a manner as possible, which we still feel is the obvious approach, given the intrinsic interrelatedness of so much of our subject matter.

While the fundamental arrangement is, of course, still intended to unify the whole work, we have tried to keep each chapter as much of a 'stand-alone' as possible, in an attempt to make this a book which permits the interested reader to just 'dip in'. Ultimately, of course, it still remains for that reader to decide how successful we have been.

The text falls into three main parts. The early chapters again examine issues of the role and market for biotechnology in an environmental context, the essential biochemistry and microbiology which enables them to be met, and the fundamental themes of biological intervention. The technologies and applications themselves make up the central core of the book, both literally and figuratively and, fittingly, this is the largest part. Finally, some of the current aspects of, and future potential for, integration in the wider field of environmental biotechnology are discussed. There is, however, one departure from the original – this time there is no Chapter 11; no final discussion of 'The Way Ahead'. Ten years ago, environmental biotechnology was a much younger field and those predictions had some purpose. Today, it has assumed its rightful place as a realistic alternative to many of the earlier established approaches for manufacturing, land remediation, pollution control and waste management and the pace of change is now just too fast to make useful forecasts that will be meaningful for the next ten years. It would be presumptuous and in any case, on balance, we feel we got most of our forecasts about right last time; we are not about to push our luck!

Despite the passage of time and all the attempts at the rationalisation of global environmental regulation, the whole subject remains inherently context dependent – a point which inevitably recurs throughout the discussion – and local modalities can conspire to shape individual best practice in a way unknown in other branches of biotechnology. What works in one country may not in another, not because the technology is flawed, but often simply because economic, legislative or societal barriers so dictate. The environmental biotechnologist must still sometimes perform the mental equivalent of a circus act in balancing these many and different considerations. It is only to be expected, then, that the choices we have made as to what to include, and the relative importance afforded them, reflect these experiences. Some readers will take issue with those decisions, but that has always been the lot of writers. It would be unrealistic to expect that we should be treated any differently.

As we wrote in the first 'Preface', it has been said that the greatest thing that anyone can do is to make a difference. It remains our hope that with this second edition, we can again in some small way, do just that.

Acknowledgements

Writing any book always involves more people than the authors, or those who work for or on behalf of the publishing company, sometimes very directly, sometimes rather less obviously so. Remembering to say 'thank you' to those who have done something very concrete or obvious is seldom a problem and there are some old friends amongst that group – especially Rob Heap and Bob Talbott – along with Bob Rust, Graham Tebbitt, Vanessa Trescott and Bob Knight, who helped us get everything straight and in time for our first deadline all those years ago! The same thing is true of people who lend you space on their coffee table when you simply have to finish a chapter – so 'thanks' again to Linda Ormiston, OBE. Sadly one of our biggest supporters at Durham – Professor Peter Evans (no relation, by the way) – died shortly before the first edition was completed and it is our great regret that he, who had given us so much encouragement to build up the Environmental Biotechnology course and was so sympathetic to the wider objectives of this book, did not see it published.

Though the personnel has changed – Keily Larkins, Lyn Roberts and Laura Stockton last time, Liz Renwick, Rachael Ballard, Izzy Canning and Fiona Woods this – the good folk at John Wiley & Sons Ltd have been their usual unflappable selves, checking in periodically to make sure that everything is still going to plan and the final manuscript is going to come in pretty much on time.

We are of course, constantly reminded of Newton's words – that we stand on the shoulders of giants – and happily acknowledge the broader debt that we owe not only to the great biologists, biochemists and engineers, but also to all those have travelled this route before us, to our own teachers who inspired us, to our contemporaries who spurred us on and to our parents without whom, in the most literal of senses, none of this would have been possible.

We were and are deeply grateful to all of these people for their help and support – and to anyone we have missed out, we are truly sorry; the slight really was not intentional. Finally, ten years on, our dogs have changed too; Mungo and Megan are, sadly, no more; the burden of missed walks and late meals has fallen on new paws. To Bess and Nell, we can only apologise – such is the lot of the writer's mutt.

1

Introduction to Environmental Biotechnology

The Organisation for Economic Co-operation and Development (OECD) defines biotechnology as 'the application of science and technology to living organisms, as well as parts, products and models thereof, to alter living or non-living materials for the production of knowledge, goods and services' (OECD, 2002). Despite the inclusiveness of this definition, there was a time when the biotechnology sector was seen as largely medical or pharmaceutical in nature, particularly amongst the general public. While to some extent the huge research budgets of the drug companies and the widespread familiarity of their products made this viewpoint understandable, it somewhat unfairly distorted the picture. Thus therapeutic instruments were left forming the 'acceptable' face of biotechnology, while elsewhere, the science was all too frequently linked with an uneasy feeling of unnatural interference. The agricultural, industrial and environmental applications of biotechnology are potentially enormous, but the shadow of Frankenstein has often been cast across them. Genetic engineering may be relatively commonplace in pharmaceutical thinking and yet when its wider use is mooted in other spheres, such as agriculture, for example even today much of society views the possibility with suspicion, if not outright hostility.

The history of human achievement has always been episodic. For a while, one particular field of endeavour seems to hold sway as the preserve of genius and development, before the focus shifts and the next wave of progress forges ahead in a dizzy exponential rush in some entirely new direction. So it was with art in the Renaissance, music in the eighteenth century, engineering in the nineteenth and physics in the twentieth. Now it is the age of the biological – in many ways forming a kind of rebirth, following on from the heyday of the great Victorian naturalists, who provided so much input into the developing science.

It is then, perhaps, no surprise that the European Federation of Biotechnology begins its 'Brief History' of the science in the year 1859, with the publication of *On the Origin of Species by Means of Natural Selection* by Charles Darwin. Though his famous voyage aboard *HMS Beagle*, which led directly to the formulation of his (then) revolutionary ideas, took place when he was a young man, he had delayed making them known until 1858, when he made a joint

Environmental Biotechnology: Theory and Application, Second Edition Gareth M. Evans and Judith C. Furlong
© 2011 John Wiley & Sons, Ltd.

presentation before the Linnaean Society with Alfred Russell Wallace, who had, himself, independently come to very similar conclusions. Their contribution was to view evolution as the driving force of life, with successive selective pressures over time endowing living beings with optimised characteristics for survival. Neo-Darwinian thought sees the interplay of mutation and natural selection as fundamental. The irony is that Darwin himself rejected mutation as too delete- rious to be of value, seeing such organisms, in the language of the times, as 'sports' – oddities of no species benefit. Indeed, there is considerable evidence to suggest that he seems to have espoused a more Lamarckist view of biological progression, in which physical changes in an organism's lifetime were thought to shape future generations.

Darwin died in 1882. Ninety-nine years later, the first patent for a genetically modified organism was granted to Ananda Chakrabarty of the US General Elec- tric, relating to a strain of *Pseudomonas aeruginosa* engineered to express the genes for certain enzymes in order to metabolise crude oil. Twenty years on from that, the first working draft of the human genome sequence was published and the full genetic blueprint of the fruit fly, *Drosophila melanogaster*, that archetype of eukaryotic genetics research, announced – and developments have continued on what sometimes feels like an almost daily basis since then. Today biotechnology has blossomed into a major growth industry with increasing numbers of com- panies listed on the world's stock exchanges and environmental biotechnology is coming firmly into its own alongside a raft of 'clean technologies' working towards ensuring the sustainable future of our species and our planet.

Thus, at the other end of the biotech timeline, a century and a half on from *Origin of Species*, the principles it first set out remain of direct relevance, although increasingly in ways that Darwin himself could not possibly have foreseen.

The Role of Environmental Biotechnology

If pharmaceutical biotechnology represents the glamorous end of the market, then environmental applications are decidedly more in the Cinderella mould. The reasons for this are fairly obvious. The prospect of a cure for the many diseases and conditions currently promised by gene therapy and other biotech-oriented medical miracles can potentially touch us all. Our lives may, quite literally, be changed. Environmental biotechnology, by contrast, deals with far less apparently dramatic topics and, though their importance, albeit different, may be every bit as great, their direct relevance is far less readily appreciated by the bulk of the population. Cleaning up contamination and dealing rationally with wastes is, of course, in everybody's best interests, but for most people, this is simply addressing a problem which they would rather had not existed in the first place. Even for industry, though the benefits may be noticeable on the balance sheet, the likes of effluent treatment or pollution control are more of an inevitable obligation than a primary goal in themselves. In general, such activities are typically funded on a distinctly limited budget and have traditionally been viewed as a necessary inconvenience. This is in no way intended to be disparaging to industry; it simply represents commercial reality.

In many respects, there is a logical fit between this thinking and the aims of environmental biotechnology. For all the media circus surrounding the grand questions of our age, it is easy to forget that not all forms of biotechnology involve xenotransplantation, genetic modification, the use of stem cells or cloning. Some of the potentially most beneficial uses of biological engineering, and which may touch the lives of the majority of people, however indirectly, involve much simpler approaches. Less radical and showy, certainly, but powerful tools, just the same. Environmental biotechnology is fundamentally rooted in waste, in its various guises, typically being concerned with the remediation of contamination caused by previous use, the impact reduction of current activity or the control of pollution. Thus, the principal aims of this field are the manufacture of products in environmentally harmonious ways, which allow for the minimisation of harmful solids, liquids or gaseous outputs or the clean-up of the residual effects of earlier human occupation.

The means by which this may be achieved are essentially twofold. Environmental biotechnologists may enhance or optimise conditions for existing biological systems to make their activities happen faster or more efficiently, or they resort to some form of alteration to bring about the desired outcome. The variety of organisms which may play a part in environmental applications of biotechnology is huge, ranging from microbes through to trees and all are utilised on one of the same three fundamental bases – accept, acclimatise or alter. For the vast majority of cases, it is the former approach, accepting and making use of existing species in their natural, unmodified form, which predominates.

The Scope for Use

There are three key points for environmental biotechnology interventions, namely in the manufacturing process, waste management or pollution control, as shown in Figure 1.1.

Accordingly, the range of businesses to which environmental biotechnology has potential relevance is almost limitless. One area where this is most apparent

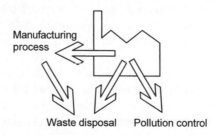

Manufacturing process

Waste disposal   Pollution control

Figure 1.1   *The three intervention points*

is with regard to waste. All commercial operations generate waste of one form or another and for many, a proportion of what is produced is biodegradable. With disposal costs rising steadily across the world, dealing with refuse constitutes an increasingly high contribution to overheads. Thus, there is a clear incentive for all businesses to identify potentially cost-cutting approaches to waste and employ them where possible. Changes in legislation throughout Europe, the US and elsewhere, combined with growing environmental awareness and a burgeoning demand for reduced carbon footprints have inevitably driven these issues higher up the political agenda and biological methods of waste treatment have gained far greater acceptance as a result. For those industries with particularly high biowaste production, the various available treatment biotechnologies can offer considerable savings.

Manufacturing industries can benefit from the applications of whole organisms or isolated bio-components. Compared with conventional chemical processes, microbes and enzymes typically function at lower temperatures and pressures. The lower energy demands this makes leads to reduced costs, but also has clear benefits in terms of both the environment and work-place safety. Additionally, biotechnology can be of further commercial significance by converting low cost organic feedstocks into high value products or, since enzymatic reactions are more highly specific than their chemical counterparts, by deriving final substances of high relative purity. Almost inevitably, manufacturing companies produce wastewaters or effluents, many of which contain biodegradable contaminants, in varying degrees. Though traditional permitted discharges to sewer or watercourses may be adequate for some, other industries, particularly those with recalcitrant or highly concentrated effluents, have found significant benefits to be gained from using biological treatment methods themselves on site. Though careful monitoring and process control are essential, biotechnology stands as a particularly cost-effective means of reducing the pollution potential of wastewater, leading to enhanced public relations, compliance with environmental legislation and quantifiable cost-savings to the business.

Those involved in processing organic matter, for example or with drying, printing, painting or coating processes, may give rise to the release of volatile organic compounds (VOCs) or odours, both of which represent environmental nuisances, though the former is more damaging than the latter. For many, it is not possible to avoid producing these emissions altogether, which leaves treating them to remove the offending contaminants the only practical solution. Especially for relatively low concentrations of readily water soluble VOCs or odorous chemicals, biological technologies can offer an economic and effective alternative to conventional methods.

The use of biological cleaning agents is another area of potential benefit, especially where there is a need to remove oils and fats from process equipment, work surfaces or drains. Aside from typically reducing energy costs, this may also obviate the need for toxic or dangerous chemical agents. The pharmaceutical and brewing industries, for example both have a long history of employing enzyme-based cleaners to remove organic residues from their process equipment.

In addition, the development of effective biosensors, powerful tools which rely on biochemical reactions to detect specific substances, has brought benefits to a wide range of sectors, including the manufacturing, engineering, chemical, water, food and beverage industries. With their ability to detect even small amounts of their particular target chemicals, quickly, easily and accurately, they have been enthusiastically adopted for a variety of process monitoring applications, particularly in respect of pollution assessment and control.

Contaminated land is a growing concern for the construction industry, as it seeks to balance the need for more houses and offices with wider social and environmental goals. The re-use of former industrial sites, many of which occupy prime locations, may typically have associated planning conditions attached which demand that the land be cleaned-up as part of the development process. With urban regeneration and the reclamation of 'brown-field' sites increasingly favoured in many countries over the use of virgin land, remediation has come to play a significant role and the industry has an on-going interest in identifying cost-effective methods of achieving it. Historically, much of this has involved simply digging up the contaminated soil and removing it to landfill elsewhere. Bioremediation technologies provide a competitive and sustainable alternative and in many cases, the lower disturbance allows the overall scheme to make faster progress.

As the previous brief examples show, the range of those which may benefit from the application of biotechnology is lengthy and includes the chemical, pharmaceutical, water, waste management and leisure industries, as well as manufacturing, the military, energy generation, agriculture and horticulture. Clearly, then, this may have relevance to the viability of these ventures and, as was mentioned at the outset, biotechnology is an essentially commercial activity. Environmental biotechnology must compete in a world governed by the Best Practicable Environmental Option (BPEO) and the Best Available Techniques Not Entailing Excessive Cost (BATNEEC). Consequently, the economic aspect will always have a large influence on the uptake of all initiatives in environmental biotechnology and, most particularly, in the selection of methods to be used in any given situation. It is impossible to divorce this context from the decision-making process. By the same token, the sector itself has its own implications for the wider economy.

The Global Environmental Market

The global environmental market is undergoing a period of massive growth. In 2001, the UK's Department of Trade and Industry estimated its value at around 1500 billion US dollars, of which some 15–20% was biotech-based. Although the passage of time has now shown some of the growth forecasts then made for the following years to have been somewhat optimistic, a recent study predicts that the market will have grown to 7400 billion US dollars by 2025 (Helmut Kaiser Consultancy, 2009). There are several major factors acting as drivers for this growth, including a greater general awareness of environmental issues, the widespread adoption of sustainable best practice by industry and geo-political changes that open new territories for technology transfer. In addition, biotechnology has increasingly gained acceptance for clean manufacturing applications, with the use of biomimetics in particular showing marked expansion over recent years, while energy production, waste management and land remediation have all benefited from the ongoing trend stimulating the sales of biotechnology-based environmental processing methods. Water treatment in its broadest sense has been perhaps the biggest winner in all this, the sector now accounting for some 25% of the total global environmental market (Helmut Kaiser Consultancy, 2009).

The export of environmental technologies is now a significant contributor to the global market, which will continue to expand in the burgeoning worldwide trend towards driving economic development alongside strong ecological awareness. Although such technology transfer is likely to continue to play a major role on the global scene, it is also probable that many countries will increasingly build their own comprehensive indigenous environmental industry over the coming years, thus circumventing their dependence on innovation imports.

Over the last decade, as many predicted, the regulatory framework across the world has experienced a radical tightening, with existing legislation on environmental pollution being more rigorously enforced and more stringent compliance standards implemented. It is hard to imagine that this trend will stop in the coming years, which once again feeds the expectation that it will act as a significant stimulus for the sales of biotechnology-based environmental processing methods. This would seem particularly likely in the current global main markets for environmental technologies, namely Asia in general, China, Japan, Europe and the USA (Helmut Kaiser Consultancy, 2009).

The benefits are not, however, confined to the balance sheet. The OECD (2001) concluded that the industrial use of biotechnology commonly leads to increasingly environmentally harmonious processes and additionally results in lowered operating and/or capital costs. For years, industry has appeared locked into a seemingly unbreakable cycle of growth achieved at the cost of environmental damage. This OECD investigation provided probably the first hard evidence to support the reality of biotechnology's long heralded promise of alternative production methods which are ecologically sound and economically efficient. A variety of industrial sectors, including pharmaceuticals, chemicals, textiles, food and energy were examined, with a particular emphasis on biomass renewable resources, enzymes and bio-catalysis. While such approaches may have to be used in tandem with other processes for maximum effectiveness, it seems that their use invariably leads to reduction in operating or capital costs, or both. Moreover, the research also concluded that it is clearly in the interests of governments of the developed and developing worlds alike to promote the use of biotechnology for the substantial reductions in resource and energy consumption, emissions, pollution and waste production it offers. The potential contribution to be made by the appropriate use of biotechnology to both environmental and economic sustainability would seem to be clear.

The upshot of this is that few biotech companies in the environmental sector perceive problems for their own business development models, principally as a result of the wide range of businesses for which their services are applicable and the large potential for growth. Competition within the sector is not seen as a major issue either, since the field is still largely open and unsaturated, and from the employment perspective, the biotech industry seems a robust one. Although the economic downturn saw the UK science labour market in general shed both permanent and contract staff throughout 2009, the biotech sector increased its demand for skilled scientists and predictions suggest that it will continue to buck the trend in the future (SRG, 2010). Moreover, there has been an established tendency towards niche specificity, with companies operating in more specialised sub-arenas within the environmental biotechnology umbrella. Given the number and diversity of such possible slots, coupled with the fact that new opportunities, and the technologies to capitalise on them, are developing apace, this trend seems likely to continue, though the business landscape is beginning to change. In some sectors, aggressive rivalry for market penetration has begun to produce bigger, multi-disciplinary environmental companies, largely through partnerships, acquisitions and direct competition. It is not without some irony that companies basing their commercial activities on biological organisms should themselves come to behave in such a Darwinian fashion. However, the picture is not entirely rosy.

Typically the sector comprises a number of relatively small, specialist companies. According to the *OECD Biotechnology Statistics 2009*, based on government survey data for 22 OECD countries and an additional four non-member countries, the majority of both biotechnology and biotechnology R&D companies have fewer than 50 employees – the average by country being 67 and 63% respectively (van Beuzekom and Arundel, 2009). As a consequence, the market has tended to be somewhat fragmented. Often the complexities of individual projects make the application of 'standard' off-the-shelf approaches very difficult, inevitably meaning that much of what is done must be significantly customised. While this, of course, is a strength and of great potential environmental benefit, it also has hard commercial implications which must be taken into account. Although the situation has begun to be addressed over recent years, historically a sizeable proportion of companies active in this sphere have had few products or services which might reasonably be termed suitable for generalised use, though they may have enough expertise, experience or sufficiently perfected techniques to deal with a large number of possible scenarios.

Historically, one of the major barriers to the wider uptake of biological approaches has been the high perceived cost of these applications. For many years, the solutions to all environmental problems were seen as expensive and for some, particularly those unfamiliar with the multiplicity of varied technologies available, this view has been slow to fade. Generally, there is often a lack of financial resource allocation available for this kind of work and biotech providers have sometimes come under pressure to reduce the prices for their services as a result. Awareness of the benefits of biotechnology, both as a means to boost existing markets and for the opening up of new ones, has undoubtedly been growing over the last ten years but it remains an important area to be addressed. The lack of marketing expertise that had formed one of the principal obstacles to the exploitation of novel opportunities, particularly in the UK, has been largely overcome, while in addition, technical understanding of biotech approaches amongst many target industries has also risen. Good education, in the widest sense, of customers and potential users of biological solutions will remain a major factor in the development and furtherance of these technologies.

Modalities and local influences

Another of the key factors affecting the practical uptake of environmental biotechnology is the effect of local circumstances. Contextual sensitivity is almost certainly the single most important factor in technology selection and represents a major influence on the likely penetration of biotech processes into the market place. Neither the nature of the biological system, nor of the application method itself, play anything like so relevant a role. This may seem somewhat unexpected at first sight, but the reasons for it are obvious on further inspection. While the character of both the specific organisms and the engineering remain essentially the same irrespective of location, external modalities of economics, legislation and custom vary on exactly this basis. Accordingly, what may make abundant sense as a biotech intervention in one region or country, may be totally unsuited to use in another. In as much as it is impossible to discount the wider global economic aspects in the discussion, disassociating political, fiscal and social conditions equally cannot be done, as the following example illustrates. Back in 1994, the expense of bioremediating contaminated soil in the United Kingdom greatly exceeded the cost of removing it to landfill. Within six years successive changes of legislation and the imposition of a landfill tax, the situation was almost completely reversed. Unsurprisingly, in those countries where landfill had always been an expensive option and thus played less of a major role in national waste management strategy, remediation has generally tended to be embraced far more readily.

While it is inevitable that environmental biotechnology must be considered as contextually dependent, clearly as the previous example shows, those contexts can change. In the final analysis, it is often fiscal instruments, rather than the technologies, which provide the driving force and sometimes seemingly minor modifications in apparently unrelated sectors can have major ramifications for the application of biotechnology. Again as has been discussed, the legal framework is another aspect of undeniable importance in this respect. Increasingly tough environmental law makes a significant contribution to the sector and changes in regulatory legislation are often enormously influential in boosting existing markets or creating new ones. When legislation and economic pressure combine, as, for example they have with the likes of European Directives on Landfill, Integrated Pollution Control and Urban Wastewater, the impetus towards a fundamental paradigm shift becomes overwhelming and the implications for relevant biological applications can be immense.

There is a natural tendency to delineate, seeking to characterise technologies into particular categories or divisions. However, the essence of environmental biotechnology is such that there are many more similarities than differences. Though it is, of course, often helpful to view individual technology uses as distinct, particularly when considering treatment options for a given environmental problem, there are inevitably recurrent themes which feature throughout the whole topic. Moreover, this is a truly applied science. While the importance of the laboratory bench cannot be denied, the controlled world of research translates imperfectly into the harsh realities of commercial implementation. Thus, there can often be a dichotomy between theory and application and it is precisely this fertile ground which is explored in the present work. In addition, the principal underlying approach of specifically *environmental* biotechnology, as distinct from other kinds, is the reliance on existing natural cycles, often directly and in an entirely unmodified form. Thus, this science stands on a foundation of fundamental biology and biochemistry. To understand the application, the biotechnologist must simply examine the essential elements of life, living systems and ecological circulation sequences. However engineered the approach, this fact remains unwaveringly true. In many respects, environmental biotechnology stands as the purest example of the multi-faceted bio-industry, since it is the least refined, at least in terms of the basis of its action. In essence, all of its applications simply encourage the natural propensity of the organisms involved, while seeking to enhance or accelerate their action. Hence, optimisation, rather than modification, is the typical route by which the particular desired end result, whatever it may be, is achieved and, consequently, a number of issues feature as common threads within the discussions of individual technologies.

Integrated Approach

Integration is an important aspect for environmental biotechnology. One theme that will be developed throughout this book is the potential for different biological approaches to be combined within treatment trains, thereby producing an overall effect which would be impossible for any single technology alone to achieve. However, the wider goal of integration is not, of necessity, confined solely to the specific methods used. It applies equally to the underpinning knowledge that enables them to function in the first place and an understanding of this is central to the rationale behind this book. In some spheres, traditional biology has become rather unfashionable and the emphasis has shifted to more exciting sounding aspects of life science. While the new-found concentration on 'ecological processes', or whatever, sounds distinctly more 'environmental', in many ways, and somewhat paradoxically, it sometimes serves the needs of environmental biotechnology rather less well. The fundamentals of living systems are the stuff of this branch of science and complex though the whole picture may be, at its simplest the environmental biotechnologist is principally concerned with a relatively small number of basic cycles. In this respect, a good working knowledge of biological processes like respiration, fermentation and photosynthesis, a grasp of the major cycles by which carbon, nitrogen and water are recycled and an appreciation of the flow of energy through the biosphere must be viewed as prerequisites. Unsurprisingly, then, these basic processes appear throughout this book, either explicitly or tacitly accepted as underpinning the context of the discussion. The intent here has been neither to insult the readership by parading what is already well known, nor gloss over aspects which, if left unexplained, at least in reasonable detail, might only serve to confuse. However, this is expressly not designed to be a substitute for much more specific texts on these subjects, nor an entire alternative to a cohesive course on biology or biochemistry. The intention is to introduce and explain the necessary aspects and elements of various metabolic pathways, reactions and abilities as required to advance the reader's understanding of this particular branch of biotechnology.

A large part of the reasons for approaching the subject in this way is the fact that there really is no such thing as a 'typical' environmental biotechnologist any more than there is a 'typical' environmental science student. The qualifications, knowledge base and experience of each means that practitioners come into the profession from a wide variety of disciplines and by many different routes. Thus, amongst their ranks are agronomists, biochemists, biologists, botanists, enzymologists, geneticists, microbiologists, molecular biologists, process engineers and protein technologists, all of whom bring their own particular skills, knowledge base and experiences. The applied nature of environmental biotechnology is obvious. While the science underlying the processes themselves may be as pure as any other, what distinguishes this branch of biological technology are the distinctly real-life purposes to which it is put. Hence, part of the intended function of this book is to attempt to elucidate the former in order to establish the basis of the latter. At the same time, as any applied scientist will confirm, what happens in the field under operational conditions represents a distinct compromise between the theoretical and the practically achievable. At times, anything more than an approximation to the expected results may be counted as something of a triumph of environmental engineering.

Closing Remarks

The celebrated astronomer and biologist, Sir Fred Hoyle, said that the solutions to major unresolved problems should be sought by the exploration of radical hypotheses, while simultaneously adhering to well tried and tested scientific tools and methods. This approach is particularly valid for environmental biotechnology. With new developments in treatment technologies appearing all the time, the list of what can be processed or remediated by biological means is ever changing. By the same token, the applications for which biotechnological solutions are sought are also subject to alteration. For the biotech sector to keep abreast of these new demands it may be necessary to examine some truly 'radical hypotheses' and possibly make use of organisms or their derivatives in ways previously unimagined. This is the basis of innovation; the inventiveness of an industry is often a good measure of its adaptability and commercial robustness.

References

van Beuzekom, B. and Arundel, A. (2009) OECD Biotechnology Statistics 2009, OECD, Paris.

Helmut Kaiser Consultancy (2009) Environmental Business Worldwide to 2025, Tübingen.

Organisation for Economic Co-operation and Development (2001) The Application of Biotechnology to Industrial Sustainability, OECD, Paris.

Organisation for Economic Co-operation and Development (2002) Frascati Manual: Proposed Standard Practice for Surveys on Research and Development, OECD, Paris.

SRG (2010) Science Labour Market Trends 2010, Slough.

2

Microbes and Metabolism

So fundamental are the concepts of cell growth and metabolic capability to the whole of environmental biotechnology and especially to remediation, that this chapter is dedicated to their exploration. Metabolic pathways (Michal, 1992) are interlinked to produce what can develop into an extraordinarily complicated network, involving several levels of control. However, they are fundamentally about the interaction of natural cycles and represent the biological element of the natural geo-biological cycles. These impinge on all aspects of the environment, both living and non-living. Using the carbon cycle as an example, carbon dioxide in the atmosphere is returned by dissolution in rainwater, and also by the process of photosynthesis to produce sugars, which are eventually metabolised to liberate the carbon once more. In addition to constant recycling through metabolic pathways, carbon is also sequestered in living and non-living components such as in trees in the relatively short term, and deep ocean systems or ancient deposits, such as carbonaceous rocks, in the long term. Cycles which involve similar principles of incorporation into biological molecules and subsequent re-release into the environment operate for nitrogen, phosphorus and sulphur. All of these overlap in some way, to produce the metabolic pathways responsible for the synthesis and degradation of bio-molecules. Superimposed, is an energy cycle, ultimately driven by the sun, and involving constant consumption and release of metabolic energy.

To appreciate the biochemical basis and underlying genetics of environmental biotechnology, at least an elementary grasp of molecular biology is required. For the benefit of readers unfamiliar with these disciplines, background information is incorporated in appropriate figures.

The Immobilisation, Degradation or Monitoring of Pollutants from a Biological Origin

Removal of a material from an environment takes one of two routes: it is either degraded or immobilised by a process which renders it biologically unavailable for degradation and so is effectively removed.

Immobilisation can be achieved by chemicals excreted by an organism or by chemicals in the neighbouring environment which trap or chelate a molecule thus making it insoluble. Since virtually all biological processes require the substrate

Environmental Biotechnology: Theory and Application, Second Edition Gareth M. Evans and Judith C. Furlong
© 2011 John Wiley & Sons, Ltd.

to be dissolved in water, chelation renders the substance unavailable. In some instances this is a desirable end result and may be viewed as a form of remediation, since it stabilises the contaminant. In other cases it is a nuisance, as digestion would be the preferable option. Such 'unwanted' immobilisation can be a major problem in remediation, and is a common state of affairs with aged contamination. Much research effort is being applied to find methods to reverse the process.

Degradation is achieved by metabolic pathways operating within an organism or combination of organisms, sometimes described as consortia. These processes are the crux of environmental biotechnology and thus form the major part of this chapter. Enzyme activity operates either through metabolic pathways functioning within the cell, or excreted by the cell or, isolated and applied in a purified form.

Biological monitoring utilises proteins, of which enzymes are a subset, produced by cells, usually to identify, or quantify contaminants. This has been developed into an expanding field of biosensor production.

Who are the biological players in these processes, what are their attributes which are so essential to this science and which types of biological material are being addressed here? The answers to these questions lie throughout this book and are summarised in this chapter.

The players

Traditionally, life was placed into two categories – those having a true *nucleus* (eukaryotes) and those that do not (prokaryotes). This view was dramatically disturbed in 1977 when Carl Woese proposed a third domain, the archaebacteria, now described as archaea, arguing that although apparently prokaryote at first glance they contain sufficient similarities with eukaryotes, in addition to unique features of their own, to merit their own classification (Woese and Fox, 1977; Woese, Kandler and Wheelis 1990). The arguments raised by this proposal continue (Cavalier-Smith, 2002) but throughout this book the classification adopted is that of Woese, namely, that there are three divisions: bacteria, archaea (which together comprise prokaryotes) and eukaryotes. By this definition, then, what are referred to throughout this work simply as 'bacteria' are synonymous with the term eubacteria (meaning 'true' bacteria).

It is primarily to the archaea, which typically inhabit extreme niches with respect to temperature, pressure, salt concentration or osmotic pressure, that a great debt of gratitude is owed for providing this planet with the metabolic capability to carry out processes under some very odd conditions indeed. The importance to environmental biotechnology of life in extreme environments is addressed in Chapter 3.

An appreciation of the existence of these classifications is important, as they differ from each other in the detail of their cell organisation and cellular processes making it unlikely that their genes are directly interchangeable. The relevance of this becomes obvious when genetic engineering is discussed later in this book in Chapter 9. However, it is interesting to examine the potentially prokaryotic origins of the eukaryotic cell. There are many theories but the one which appears to have the most adherents is the endosymbiotic theory. It suggests that the 'proto' eukaryotic cell lost its cell wall, leaving only a membrane, and phagocytosed or subsumed various other bacteria with which it developed a symbiotic relationship. These included an aerobic bacterium, which became a mitochondrion, endowing the cell with the ability to carry out oxidative phosphorylation, a method of producing chemical energy able to be transferred to the location in the cell where it is required. Similarly, the chloroplast, the site of photosynthesis in higher plants, is thought to have been derived from cyanobacteria, the so-called blue-green algae. Chloroplasts are a type of plastid. These are membrane bound structures found in vascular plants. Far from being isolated cellular organelles, the plastids communicate with each other through interconnecting tubules (Köhler *et al.*, 1997). Various other cellular appendages are also thought to have prokaryotic origins such as cilia or the flagellum on a motile eukaryotic cell which may have formed from the fusion of a spirochete bacterium to this 'proto' eukaryote. Nuclei may well have similar origins but the evidence is still awaited.

No form of life should be overlooked as having a potential part to play in environmental biotechnology. However, the organisms most commonly discussed in this context are microbes and certain plants. They are implicated either because they are present by virtue of being in their natural environment or by deliberate introduction.

Microbes

Microbes are referred to as such, simply because they can not be seen by the naked eye. Many are bacteria or archaea, all of which are prokaryotes, but the term 'microbe' also encompasses some eukaryotes, including yeasts, which are unicellular fungi, as well as protozoa and unicellular plants. In addition, there are some microscopic multicellular organisms, such as rotifers, which have an essential role to play in the microsystem ecology of places such as sewage treatment plants. Individual cells of a eukaryotic multicellular organism like a higher plant or animal, are approximately $20\,\mu m$ in diameter, while a yeast cell, also eukaryotic but unicellular, is about $5\,\mu m$ in diameter. Although bacterial cells occur in a variety of shapes and sizes, depending on the species, typically a bacterial cell is rod shaped, measuring approximately $1\,\mu m$ in width and $2\,\mu m$ in length. At its simplest visualisation, a cell, be it a unicellular organism, or one cell in a multicellular organism, is a bag, bounded by a membrane, containing an aqueous solution in which are all the molecules and structures required to enable its continued survival. In fact, this 'bag' represents a complicated infrastructure differing distinctly between prokaryotes and eukaryotes (Cavalier-Smith, 2002), but a discussion of this is beyond the scope of this book.

Depending on the microbe, a variety of other structures may be present, for instance, a cell wall providing additional protection or support, or a flagellum, a flexible tail, giving mobility through the surrounding environment. Survival requires cell growth, replication of the deoxyribonucleic acid (DNA) and then division, usually sharing the contents into two equal daughter cells. Under ideal conditions of environment and food supply, division of some bacteria may occur every 20 minutes, but most take rather longer. However, the result of many rounds of the binary division just described, is a colony of identical cells. This may be several millimetres across and can be seen clearly as a contamination on a solid surface, or if in a liquid, it will give the solution a cloudy appearance. Other forms of replication include budding off, as in some forms of yeast, or the formation of spores as in other forms of yeast and some bacteria. This is a type of DNA storage particularly resistant to environmental excesses of heat and pH, for example. When the environment becomes more hospitable, the spore can develop into a bacterium or yeast, according to its origins, and the life cycle continues.

Micro organisms may live as free individuals or as communities, either as a clone of one organism, or as a mixed group. Biofilms are examples of microbial communities, the components of which may number several hundred species. This is a fairly loose term used to describe any aggregation of microbes which coats a surface, consequently, biofilms are ubiquitous. They are of particular interest in environmental biotechnology since they represent the structure of microbial activity in many relevant technologies such as trickling filters. Models for their organisation have been proposed (Kreft *et al.*, 2001). Their structure, and interaction between their members, was of sufficient interest to warrant a major Symposium ten years ago (Allison *et al.*, 2000), and several since. Commonly, biofilms occur at a solid/liquid interphase. Here, a mixed population of microbes live in close proximity which may be mutually beneficial. Such consortia can increase the habitat range, and the overall tolerance to stress and metabolic diversity of individual members of the group. It is often thanks to such communities, rather than isolated bacterial species, that recalcitrant pollutants are eventually degraded due to combined contributions of several of its members.

Another consequence of this close proximity is the increased likelihood of bacterial transformation. This is a procedure whereby a bacterium may absorb free DNA, the macromolecule which stores genetic material, from its surroundings released by other organisms, as a result of cell death, for example. The process is dependent on the ability, or competence, of a cell to take up DNA, and upon the concentration of DNA in the surrounding environment. This is commonly referred to as horizontal transfer as opposed to vertical transfer which refers to inherited genetic material, either by sexual or asexual reproduction. Some bacteria are naturally competent, others exude competence factors and there is laboratory evidence that lightning can impart competence to some bacteria (Demaneche *et al.*, 2001). It is conceivable that conditions allowing transformation, prevail in biofilms considering the very high local concentration of microbes. Indeed there is evidence that such horizontal transfer of DNA occurs between organisms in these communities (Ehlers, 2000). In addition to transformation, genes are readily transferred on plasmids as described later in this chapter. It is now well established that, by one method or another, there is so much exchange of genetic material between bacteria in soil or in aquatic environments, that rather than discrete units, they represent a massive gene pool (Whittam, 1992).

The sliminess often associated with biofilms is usually attributed to excreted molecules usually protein and carbohydrate in nature, which may coat and protect the film. Once established, the biofilm may proliferate at a rate to cause areas of anoxia at the furthest point from the source of oxygen, thus encouraging the growth of anaerobes. Consequently, the composition of the biofilm community is likely to change with time.

To complete the picture of microbial communities, it must be appreciated that they can include the other micro organisms listed above, namely, yeasts, protozoa, unicellular plants and some microscopic multicellular organisms such as rotifers.

Plants

In contrast with microbes, the role of plants in environmental biotechnology is generally a structural one, exerting their effect by oxygenation of a microbe rich environment, filtration, solid to gas conversion or extraction of the contaminant. These examples are examined in detail in Chapters 7 and 10. Genetic modification of crop plants to produce improved or novel varieties is discussed in Chapter 9. This field of research is vast and so the discussion is confined to relevant issues in environmental biotechnology rather than biotechnology in general.

Metabolism

The energy required to carry out all cellular processes is obtained from ingested food in the case of chemotrophic cells, additionally from light in the case of phototrophs and from inorganic chemicals in lithotrophic organisms. Since all biological macromolecules contain the element carbon, a dietary source of carbon is a requirement. Ingested food is therefore, at the very least, a source of energy and carbon, the chemical form of which is rearranged by passage through various routes called metabolic pathways and the study of this flux is now sometimes referred to as metabolomics. One purpose of this reshuffling is to produce, after addition or removal of other elements such as hydrogen, oxygen, nitrogen, phosphorous and sulphur, all the chemicals necessary for growth. The other is to produce chemical energy in the form of adenosine triphosphate (ATP), also one of the 'building blocks' of nucleic acids. Where an organism is unable to synthesise all its dietary requirements, it must ingest them, as they are, by definition, essential nutrients. The profile of these can be diagnostic for that organism and may be used in its identification in the laboratory. An understanding of nutritional requirements of any given microbe, can prove essential for successful remediation by bioenhancement.

At the core of metabolism are the central metabolic pathways of glycolysis and the tricarboxylic acid (TCA) cycle on which a vast array of metabolic pathways eventually converge or from which they diverge. Glycolysis is the conversion of the six carbon phosphorylated sugar, glucose 6-phosphate, to the three carbon organic acid, pyruvic acid, and can be viewed as pivotal in central metabolism since from this point, pyruvate may enter various pathways determined by the energy and synthetic needs of the cell at that time. A related pathway, sharing some but not all of the reactions of glycolysis, and which operates in the opposite direction is called gluconeogenesis. Pyruvate can continue into the TCA cycle whose main function is to produce and receive metabolic intermediates and to produce energy, or into one of the many fermentation routes.

The principles of glycolysis are universal to all organisms known to date, although the detail differs between species. An outline of glycolysis, the TCA, and its close relative the glyoxalate, cycles is given in Figure 2.1, together with an indication of the key points at which the products of macromolecule catabolism, or break-down, enter these central metabolic pathways. The focus is on degradation rather than metabolism in general, since this is the crux of bioremediation. A description of the biological macromolecules which are lipids, carbohydrates, nucleic acids and proteins are given in the appropriate Figures 2.2–2.5.

Not all possible metabolic routes are present in the genome of any one organism. Those present are the result of evolution, principally of the enzymes which catalyse the various steps, and the elements which control their expression. However, an organism may have the DNA sequences, and so have

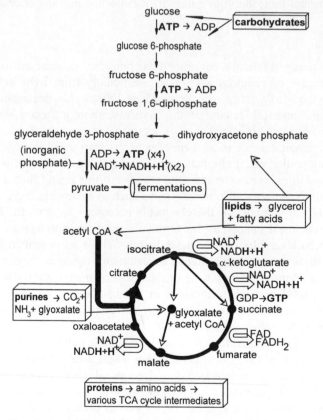

Figure 2.1   *Glycolysis, the TCA and glyoxalate cycles*

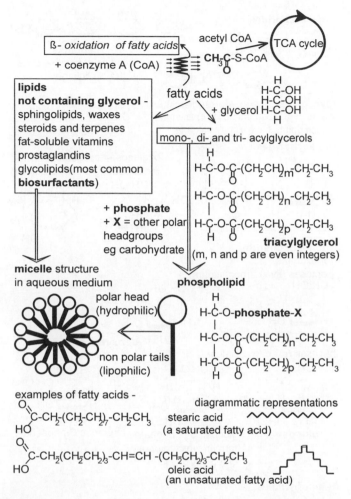

Figure 2.2 Lipids

the genetic capability, for a metabolic route even though it is not 'switched on'. This is the basis for the description of 'latent pathways' which suggests the availability of a route able to be activated when the need arises, such as challenge from a novel chemical in the environment. Additionally, there is huge potential for uptake and exchange of genetic information as discussed earlier in this chapter. It is the enormous range of metabolic capability which is harnessed in environmental biotechnology.

The basis of this discipline is about ensuring that suitable organisms are present which have the capability to perform the task required of them. This demands the provision of optimal conditions for growth, thus maximising degradation or removal of the contaminant. Linked to many of the catalytic steps in the metabolic pathway are reactions which release sufficient energy to allow the synthesis of

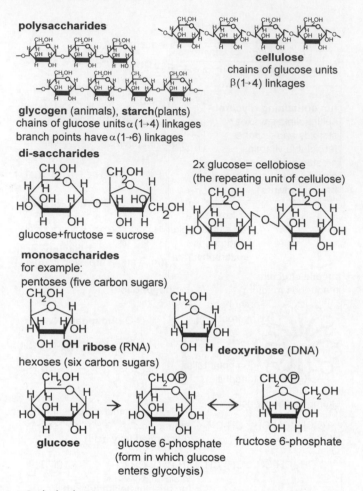

Figure 2.3   *Carbohydrates*

ATP. This is the energy 'currency' of a cell which permits the transfer of energy produced during degradation of a food to a process which may be occurring in a distant location and which requires energy.

For brevity, the discussions in this chapter consider the metabolic processes of prokaryotes and unicellular eukaryotes as equivalent to a single cell of a multi-cellular organism such as an animal or plant. This is a hideous oversimplification but justified when the points being made are general to all forms of life. Major differences are noted.

The genetic blueprint for metabolic capability

Metabolic capability is the ability of an organism or cell to digest available food. Obviously, the first requirement is that the food should be able to enter

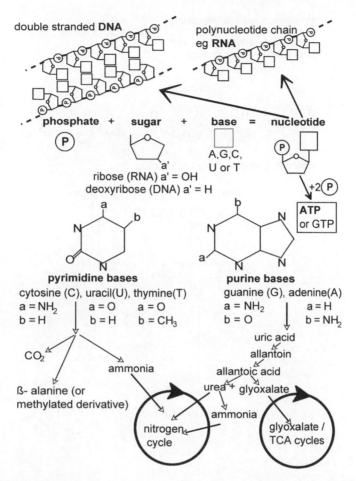

Figure 2.4   *Nucleic acids*

the cell which sometimes requires specific carrier proteins to allow penetration across the cell membrane. Once entered, the enzymes must be present to catalyse all the reactions in the pathway responsible for degradation, or catabolism. The information for this metabolic capability is encoded in the DNA. The full genetic information is described as the genome and can be a single circular piece of DNA as in bacteria, or may be linear and fragmented into chromosomes as in higher animals and plants.

Additionally, many bacteria carry plasmids, which are much smaller pieces of DNA, also circular and self-replicating. These are vitally important in the context of environmental biotechnology in that they frequently carry the genes for degradative pathways. Many of these plasmids may move between different bacteria where they replicate, thus making the metabolic capability they carry transferable. Bacteria show great promiscuity with respect to sharing their DNA.

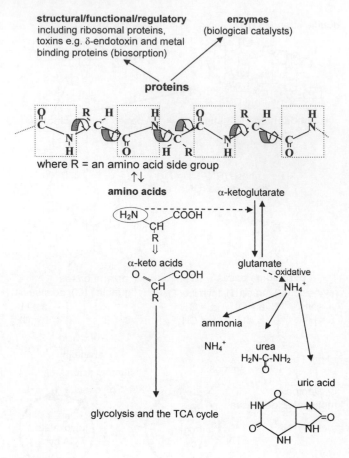

Figure 2.5   *Proteins*

Often, bacteria living in a contaminated environment, themselves develop additional degradative capabilities. The source of that genetic information new to
the organism, whether it is from modification of DNA within the organism or
transfer from other microbes, or DNA free in the environment, is a source of hot
debate between microbiologists.

DNA not only codes for RNA which is translated into proteins, but also for
RNAs which are involved in protein synthesis, namely transfer RNA (tRNA) and
ribosomal RNA (rRNA), also, small RNAs which are involved in the processing
of rRNA. These are illustrated in Figure 2.6.

There have been many systems used to describe the degree of relatedness
between organisms, but the most generally accepted is based on the sequence
of the DNA coding for rRNA, the rDNA (Stackebrandt and Woese, 1981). For
completeness, it is important to mention the retroviruses which are a group of
eukaryotic viruses with RNA rather than DNA as their genome. They carry the

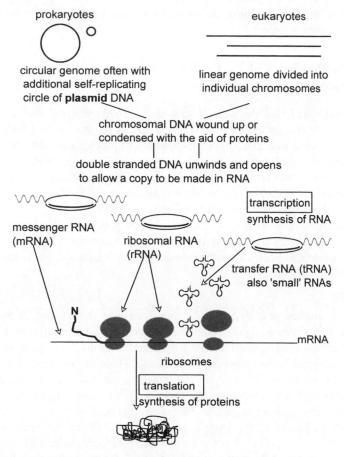

Figure 2.6 *Storage and expression of genetic information*

potential for integration into inheritable DNA due to the way in which they replicate their genomic RNA by way of a DNA intermediate.

Microbial diversity

Microbes have been discovered in extraordinarily hostile environments where their continued survival has made demands on their structure and metabolic capability. These organisms, frequently members of the archaea, are those which have the capacity to degrade some of the most hazardous and recalcitrant chemicals in our environment and thus provide a rich source of metabolic capacity to deal with some very unpleasant contaminants. This situation will remain as long as the environments which harbour these invaluable microbes are recognised as such and are not destroyed. Microbial life on this planet, taken as a whole, has an immense capability to degrade noxious contaminants; it is essential to maintain the diversity and to maximise the opportunity for microbes to metabolise the offending carbon source.

Metabolic Pathways of Particular Relevance to Environmental Biotechnology

Having established that the overall strategy of environmental biotechnology is to make use of the metabolic pathways in micro organisms to break down or metabolise organic material, this chapter now examines those pathways in some detail. Metabolic pathways operating in the overall direction of synthesis are termed anabolic while those operating in the direction of breakdown or degradation are described as catabolic: the terms catabolism and anabolism being applied to describe the degradative or synthetic processes respectively.

It has been mentioned already in this chapter and it will become clear from the forthcoming discussion, that the eventual fate of the carbon skeletons of biological macromolecules is entry into the central metabolic pathways.

Glycolysis

As the name implies, glycolysis is a process describing the splitting of a phosphate derivative of glucose, a sugar containing six carbon atoms, eventually to produce two pyruvate molecules, each having three carbon atoms. There are at least four pathways involved in the catabolism of glucose. These are the Embden-Meyerhof (Figure 2.1), which is the one most typically associated with glycolysis, the Entner-Doudoroff and the phosphoketolase pathways and the pentose phosphate cycle, which allows rearrangement into sugars containing 3, 4, 5, 6 or 7 carbon atoms. The pathways differ from each other in some of the reactions in the first half up to the point of lysis to two three-carbon molecules, after which point the remainder of the pathways are identical. These routes are characterised by the particular enzymes present in the first half of these pathways catalysing the steps between glucose and the production of dihydroxyacetone phosphate in equilibrium with glyceraldehyde 3-phosphate. All these pathways have the capacity to produce ATP and so function in the production of cellular energy. The need for four different routes for glucose catabolism, therefore, lies in the necessity for the supply of different carbon skeletons for anabolic processes and also for the provision of points of entry to glycolysis for catabolites from the vast array of functioning catabolic pathways. Not all of these pathways operate in all organisms. Even when several are encoded in the DNA, exactly which of these are active in an organism at any time, depends on its current metabolic demands and the prevailing conditions in which the microbe is living.

The point of convergence of all four pathways is at the triose phosphates which is the point where glycerol as glycerol phosphate enters glycolysis and so marks the link between catabolism of simple lipids and the central metabolic pathways. The addition of glycerol to the pool of trioses is compensated for by the action of *triose phosphate isomerase* maintaining the equilibrium between glyceraldehyde 3-phosphate and dihydroxyacetone phosphate which normally lies far in favour of the latter. This is perhaps surprising since it is glyceraldehyde 3-phosphate which is the precursor for the subsequent step. The next stage is the introduction of a second phosphate group to glyceraldehyde 3-phosphate with an accompanying oxidation, to produce glyceraldehyde 1,3-diphosphate. The oxidation involves the transfer of hydrogen to the coenzyme, NAD, to produce its reduced form, NADH. In order for glycolysis to continue operating, it is essential for the cell or organism to regenerate the NAD^+ which is achieved either by transfer of the hydrogens to the cytochromes of an electron transport chain whose operation is associated with the synthesis of ATP, or to an organic molecule such as pyruvate in which case the opportunity to synthesise ATP is lost. This latter method is the first step of many different fermentation routes. These occur when operation of electron transport chains is not possible and so become the only route for the essential regeneration of NAD^+. Looking at the Embden-Meyerhof pathway, this is also the third stage at which a phosphorylation has occurred. In this case, the phosphate was derived from an inorganic source, in a reaction which conserves the energy of oxidation.

The next step in glycolysis is to transfer the new phosphate group to ADP, thus producing ATP and 3-phosphoglycerate, which is therefore the first substrate level site of ATP synthesis. After rearrangement to 2-phosphoglycerate and dehydration to phosphoenolpyruvic acid, the second phosphate is removed to produce pyruvic acid and ATP, and so is the second site of substrate level ATP synthesis. As mentioned above, depending on the activity of the electron transport chains and the energy requirements of the cell balanced against the need for certain metabolic intermediates, pyruvate, or its derivatives may now be reduced by accepting the hydrogen from NADH and so continue on a fermentation route or it may be decarboxylated to an acetyl group and enter the TCA cycle. The overall energy balance of glycolysis is discussed later when considering chemical cellular energy production in more detail.

TCA cycle

Pyruvate decarboxylation produces the acetyl group bound to Coenzyme A, ready to enter the TCA cycle otherwise named Kreb's Citric Acid Cycle in tribute to the scientist who discovered it. Not only is this cycle a source of reduced cofactors which 'fuel' electron transport and thus, the synthesis of ATP, but it is also a great meeting point of metabolic pathways. Cycle intermediates are constantly being removed or replenished. During anaerobic fermentation, many of the reactions seen in the TCA cycle are in operation even though they are not linked to electron transport.

Glyoxalate cycle

This is principally the TCA cycle, with two additional steps forming a 'short circuit', involving the formation of glyoxalate from isocitrate. The second reaction requires the addition of acetyl CoA to glyoxalate to produce malic acid and thus rejoin the TCA cycle. The purpose of this shunt is to permit the organism to use acetyl CoA, which is the major breakdown product of fatty acids, as its sole carbon source.

Macromolecules – description and degradation

Lipids

This class of macromolecules (see Figure 2.2) includes the neutral lipids which are triacylglycerols commonly referred to as fats and oils. Triacylglycerols are found in reservoirs in micro-organisms as fat droplets, enclosed within a 'bag', called a vesicle, while in higher animals, there is dedicated adipose tissue, comprising mainly cells full of fat. These various fat stores are plundered when energy is required by the organism as the degradation of triacylglycerols is a highly exergonic reaction and therefore a ready source of cellular energy. Gram for gram, the catabolism of these fats releases much more energy than the catabolism of sugar which explains in part why energy stores are fat rather than sugar. If this were not the case the equivalent space taken up by a sugar to store the same amount of energy would be much greater. In addition, sugar is osmotically active which could present a problem for water relations within a cell, should sugar be the major energy store.

Triacylglycerols comprise a glycerol backbone onto which fatty acids are esterified to each of the three available positions. They are insoluble in an aqueous environment due to the non-polar nature of the fatty acids forming 'tails' on the triacylglycerol. However, diacylglycerols and monoacylglycerols which are esterified at only two or one position respectively, may form themselves into micelles due to their polar head, and so may exhibit apparent solubility by forming an emulsion. The tri-, di- and mono-acylglycerols have in the past been described as tri-, di- or mono-glycerides. Although these are inaccurate descriptions of the chemistry of these compounds the terms tri-, di- and mono-glycerides are still in common usage. Chemically, fats and oils are identical. If the compound in question is a liquid at room temperature, frequently it is termed an oil, if solid it is described as a fat. The melting point of these compounds is determined to a large extent by the fatty acid content, where in general, saturated fatty acids, due to their ability to pack together in an orderly manner, confer a higher melting point than unsaturated fatty acids.

Their catabolism is by hydrolysis of the fatty acids from the glycerol backbone, followed by oxidation of the fatty acids by β-oxidation. This process releases glycerol which may then be further degraded by feeding into the central pathways of glycolysis, and several units of the acetyl group attached to the carrier Coenzyme A (Figure 2.2), which may feed into the central metabolic pathways just prior to entry into the TCA cycle (Figure 2.1).

Compound lipids include the phosphoglycerides which are a major component of cell membranes. These can have very bulky polar head groups and non-polar

for example-
Rhamnolipid from *Pseudomonas*
(glycolipid)

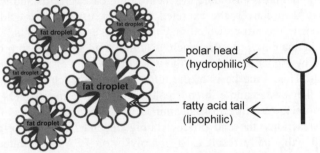

micelle structure
allowing dispersion of fats and oils in an aqueous medium

Figure 2.7   *Biosurfactants*

tails which allow them to act as surfactants and in this specific context, bio surfactants. The most common surfactants are glycolipids (Figure 2.7), which do not have a glycerol backbone, but have sugar molecules forming a polar head and fatty acids forming non-polar tails, in an overall structure similar to that shown for phospholipids in Figure 2.2. Derived lipids include fat soluble vitamins, natural rubber, cholesterol and steroid hormones. It is interesting to note here that bacteria do not synthesise steroids, and yet some, for example *Comamonas testosteroni*, are able to degrade specific members of the group; testosterone in the case given (Horinouchi *et al.*, 2001). However, oestrogen and its synthetic analogues used in the contraceptive pill are virtually recalcitrant to decomposition by bacteria. This can prove to be a problem in waterways especially in Canada where the level of such endocrine disrupters has been so high in some lakes that the feminisation of fish can be a concern (McMaster, 2001). This subject, and similar findings for the UK, are explored further in Chapter 3.

Proteins

The first catabolic step in protein degradation (see Figure 2.5) is enzymatic hydrolysis of the peptide bond formed during protein synthesis resulting in the release of short pieces, or peptides, and eventually after further degradation, amino acids. The primary step in amino acid catabolism is to remove the amino group thus producing an α-keto acid. This is usually achieved by transfer of the amino group to the TCA cycle intermediate, α-ketoglutarate, resulting in the amino acid, glutamate. Amino groups are highly conserved in all organisms due to the small number of organisms able to fix atmospheric nitrogen and so the source of an amino group is usually by transfer from another molecule. However, eventually, nitrogen is removed by oxidative deamination and is excreted in a form which depends upon the organism. Ammonia is toxic to most cells, but if an organism lives in an aqueous habitat, it may release ammonia directly into its surroundings where it is diluted and so made harmless. However, even in such an environment, if dilution should prove insufficient, ammonia concentration will increase, likewise the pH, consequently, the well-being of the organism will be compromised. Organisms which cannot make use of dilution, rid themselves of ammonia by converting it first into a less toxic form such as urea in the case of mammals and the fairly insoluble uric acid in the case of birds and most reptiles. Bacteria may then convert the excreted ammonia, urea or uric acid into nitrite and then oxidise it to nitrate which may then be taken up by plants. From there it is included in anabolic processes such as amino acid synthesis to produce material ingested by higher animals and the whole procedure of amino group transfer repeats itself. This is the basis of the nitrogen cycle which forms a central part of much of the sewage and effluent treatment described in Chapters 6 and 7.

The α-keto acid resulting from deamination of the amino acid is degraded by a series of reactions, the end product being dependent on the original amino acid, but all will finally result as a glycolysis or TCA cycle intermediate. A fascinating story of catabolism showing collaboration between mammals and bacteria resident in the gut, is the degradation of haemoglobin, the component of blood which carries oxygen and carbon dioxide. Haemoglobin comprises the protein, globin, into which was inserted during synthesis, the haem ring system where the exchange between binding of oxygen or carbon dioxide takes place in circulating blood. The first step of haemoglobin degradation, performed in the mammalian system, is removal of the haeme ring structure releasing globin which is subject to normal protein degradation. Haeme has its origins in the amino acids in that the starting point for the ring structure is the amino acid, glycine. The degradation pathways start with removal of iron and release of carbon monoxide to produce the linear structure, bilirubin. This is eventually excreted into the gut where enteric (gut) bacteria degrade the bilirubin to urobilinogens which are degraded further, some being excreted in the urine and others, such as stercobilin, are excreted in the faeces. All these products are further metabolised by microbes, for example in the sewage treatment plant.

Nucleic acids

Degradation of nucleic acids (see Figure 2.4) is also a source of ammonium ion. The purines are broken down to release CO_2 and uric acid which is reduced to allantoin. This is then hydrolysed to produce urea and glyoxylate which can enter the TCA cycle by the glyoxylate pathway present in plants and bacteria but not mammals. The urea thus produced may be further hydrolysed to ammonium ion or ammonia with the release of carbon dioxide. The form in which the nitrogen derived from the purines is excreted again depends upon the organism.

Pyrimidines are hydrolysed to produce ammonia which enters the nitrogen cycle, carbon dioxide and β-alanine or β-aminoisobutyric acid both of which are finally degraded to succinyl CoA which enters the TCA cycle.

Carbohydrates

The carbohydrates (see Figure 2.3) form a ready source of energy for most organisms as they lead, by a very short route, into the central metabolic pathways from which energy to fuel metabolic processes is derived. When several sugar units, such as glucose, are joined together to form macromolecules, they are called polysaccharides. Examples of these are glycogen in animals, and cellulose in plants. In nature, the sugars usually occur as ring structures and many have the general formula, $C(H_2O)_n$, where carbon and water are present in equal proportion. Catabolism of glucose has been described earlier in this chapter. As stated earlier, the resulting metabolite from a given carbon source, or the presence of specific enzymes, can be diagnostic of an organism. Whether or not the enzymes of a particular route are present can help to identify a microbe, and carbohydrate metabolism is frequently the basis of micro-organism identification in a public health laboratory. Glucose enters the glycolytic pathway to pyruvate, the remainder of which is determined in part by the energy requirements of the cell and in part by the availability of oxygen. If the organism or cell normally exists in an aerobic environment, there is oxygen available and the pyruvate is not required as a starting point for the synthesis of another molecule, then it is likely to enter the TCA cycle. If no oxygen is available, fermentation, defined later in this chapter, is the likely route. The function of fermentation is to balance the chemical reductions and oxidations performed in the initial stages of glycolysis.

Production of Cellular Energy

Cellular energy is present mainly in the form of ATP and to a lesser extent, GTP (Figure 2.4) which are high energy molecules, so called because a large amount of chemical energy is released on hydrolysis of the phosphate groups. The energy to make these molecules is derived from the catabolism of a food, or from photosynthesis. A food source is commonly carbohydrate, lipid or to a lesser extent, protein but if a compound considered to be a contaminant can enter a catabolic pathway, then it can become a 'food' for the organism. This is the basis of bioremediation. The way in which energy is transferred from the 'food' molecule to ATP may take two substantially different routes. One is cytoplasmic synthesis of ATP which is the direct transfer of a phosphate group to ADP, storing the energy of that reaction in chemical bonds. The other involves a fairly complicated system involving transfer of electrons and protons, or hydrogen ions, which originated from the oxidation of the 'food' at some stage during its passage through the catabolic pathways. The final sink for the electrons and hydrogen ions is oxygen, in the case of oxidative phosphorylation, to produce water. This explains the need for good aeration in many of the processes of environmental biotechnology, where organisms are using oxidative phosphorylation as their main method for synthesising ATP. An example of this is the activated sludge process in sewage treatment. However, many microbes are anaerobes, an example being a class of archaea, the methanogens, which are obligate anaerobes in that they will die if presented with an oxygenated atmosphere. This being the case, they are unable to utilise the oxidative phosphorylation pathways and so instead, operate an electron transport chain similar in principle, although not in detail. It has as the ultimate electron and hydrogen sink, a variety of simple organic compounds including acetic acid, methanol and carbon dioxide. In this case, the end product is methane in addition to carbon dioxide or water depending on the identity of the electron sink. These are the processes responsible for the production of methane in an anaerobic digester which explains the necessity to exclude air from the process.

Fermentation and respiration

The electrons derived from the catabolism of the carbon source are eventually either donated to an organic molecule in which case the process is described as fermentation, or donated to an inorganic acceptor by transfer along an electron chain. This latter process is respiration and may be aerobic where the terminal electron acceptor is oxygen, or anaerobic where the terminal electron acceptor is other than oxygen such as nitrate, sulphate, carbon dioxide, sulphur or ferric ion. Unfortunately, respiration is a term which has more than one definition. It may also be used to describe a subset of the respiration processes mentioned above to include only oxidation of organic material and where the ultimate electron acceptor is molecular oxygen. This latter definition is the basis of Biological Oxygen Demand (BOD), which is often used to characterise potential environmental pollutants, especially effluents, being a measure of the biodegradable material available for oxidation by microbes.

Fermentations

In modern parlance, there are many definitions of the term 'fermentation'. They range from the broadest and somewhat archaic to mean any large scale culture of micro organisms, to the very specific, meaning growth on an organic substance and which is wholly dependent on substrate level phosphorylation. This is the synthesis of ATP by transfer of a phosphate group directly from a high energy compound and not involving an electron transport chain. Additionally, and a source of great confusion, is that fermentation may refer simply to any microbial growth in the absence of oxygen but equally may be used generally to mean microbial growth such as food spoilage where the presence or absence of oxygen is unspecified. The definition used throughout this book, except with reference to eutrophic fermentation discussed in Chapter 8, is that of growth dependent on substrate level phosphorylation.

There are very many fermentation routes but all share two requirements, the first being the regeneration of NAD^+ from NADH produced during glycolysis which is essential to maintain the overall reduction: oxidation equilibrium, and the second being that pyruvate, or a derivative thereof, is the electron acceptor during the reoxidation of NADH. What this means is that all fermentation routes start with pyruvate, the end point of glycolysis, and proceed along a variety of pathways to an end product indicative, if not diagnostic, of the organism. Fermentation is therefore an option under conditions where there is an active electron transport chain as discussed in the following section, but becomes essential when fermentation is the only method for regenerating NAD^+.

As noted above, the end product of fermentation for any given carbon source may be diagnostic of the identity of a specific organism. This is more relevant for bacteria than for yeast or other eukaryotic cells and arises from the predisposition of that organism, to use a particular fermentation pathway. These are described in detail in Mandelstam and McQuillen (1973) and are summarised in Figure 2.8. Identification by the product of carbohydrate catabolism is somewhat specialised and is very thoroughly set out in *Cowan and Steel's Manual for the Identification of Medical Bacteria* (Barrow and Feltham, 1993).

Electron transport chains: oxidative phosphorylation and methanogenesis

As described in the previous section, NADH and other reduced cofactors may be reoxidised by the reduction of organic receptors such as pyruvate. This is the fermentation route.

Alternatively, the reducing agent (or reductant) can transfer the electrons to an electron transport chain which ultimately donates them to an inorganic receptor (the oxidising agent or oxidant). In aerobic respiration, this receptor is oxygen. However, some bacteria have electron transport chains which use other electron sinks such as nitrate, sulphate, carbon dioxide and some metals, with respiration being described as anaerobic in these cases. The use of nitrate in this role leads to the process of denitrification, which plays an important part in many aspects of the applications of environmental biotechnology.

A number of events occur during the flow of electrons along the chain which have been observed and clearly described for a number of organisms and organelles, most especially the mitochondria of eukaryotic cells. These are fully discussed in many biochemistry textbooks, an excellent example being

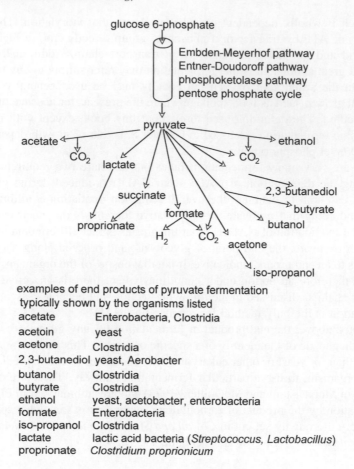

glucose 6-phosphate

Embden-Meyerhof pathway
Entner-Doudoroff pathway
phosphoketolase pathway
pentose phosphate cycle

pyruvate

acetate   ethanol
CO_2 lactate CO_2

                                    2,3-butanediol
            succinate   butyrate
                        formate
        proprionate   butanol
                H_2 CO_2
                            acetone

                            iso-propanol

examples of end products of pyruvate fermentation
typically shown by the organisms listed

acetate	Enterobacteria, Clostridia
acetoin	yeast
acetone	Clostridia
2,3-butanediol	yeast, Aerobacter
butanol	Clostridia
butyrate	Clostridia
ethanol	yeast, acetobacter, enterobacteria
formate	Enterobacteria
iso-propanol	Clostridia
lactate	lactic acid bacteria (*Streptococcus, Lactobacillus*)
proprionate	*Clostridium proprionicum*

Figure 2.8   *Fermentations*

Lehninger (1975) the gist of which is outlined in this section. The details of exactly how these phenomena combine to drive the synthesis of ATP is still unclear but various models have been proposed.

The chemiosmotic model, proposed by Peter Mitchell in 1961, states that the proton, or hydrogen ion, gradient which develops across an intact membrane during biological oxidations is the energy store for the subsequent synthesis of ATP. This model somewhat revolutionised the then current thinking on the energy source for many cellular processes, as the principles of energy storage and availability according to the chemiosmotic theory were applicable to many energy demanding cellular phenomena including photosynthetic phosphorylation and some cross-membrane transport systems. It could even account for the movement of flagellae which propel those bacteria possessing them, through a liquid medium. The chemiosmotic theory accounts for the coupling of the transmembrane proton gradient to ATP synthesis. It implies that during oxidation, the electrons flow down from high to low energy using that energy to drive protons across a membrane against a high concentration, thus developing the proton gradient. When the electron flow stops, the protons migrate down the concentration gradient, simultaneously releasing energy to drive the synthesis of ATP through membrane associated proteins. The model system described first is that of mitochondria and later in this chapter, comparisons with bacterial systems associated with oxidative phosphorylation and those systems associated with methanogenesis will be made.

Electron transport chains comprise cytochrome molecules which trap electrons, and enzymes which transfer electrons from a cytochrome to its neighbour. The quantity of energy released during this transfer is sufficient to drive the synthesis of approximately one ATP molecule by the enzyme *ATP synthetase*. The whole system is located in a membrane which is an essential requirement of any electron transport chain because of the need to organise it topographically, and to allow the establishment of a pH gradient. Also there is evidence that during active electron transport, the morphology of the membrane changes and is believed to store energy in some way yet to be fully elucidated. Consequently, an intact membrane is essential. Any toxic substance which damages the integrity of a membrane has the potential to interrupt the functioning of the electron transport chain thereby reducing the facility for ATP synthesis and potentially killing the organism. The chain may also be disrupted by interference with the electron carriers. Such a chemical is cyanide, which complexes with cytochrome oxidase, and for which research into a biological remediation route is underway.

The mitochondrial electron transport system and oxidative phosphorylation

The electron transport system in eukaryotes is located in the inner membrane of mitochondria. A representation of the system is given in Figure 2.9.

The chain is a series of complexes comprising cytochromes, and enzymes involved in oxidation-reduction reactions whose function is to transfer electrons from one complex to the next. The ratios of the complexes one to another

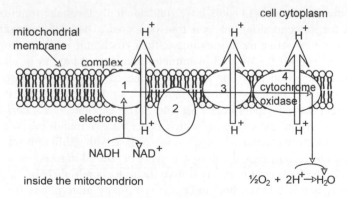

Figure 2.9 *Mitochondrial electron transport chain*

varies from cell type to cell type, however the concentration of the *cytochrome a* complex per unit area of inner membrane stays fairly constant. What changes from cell type to cell type is the degree of infolding of the inner membrane, such that cells requiring a large amount of energy have mitochondria which have a very large surface area of inner membrane, which is highly convoluted thus providing a high capacity for electron transport. The process which couples ATP synthesis to electron transport in mitochondria and which still evades a complete description, is oxidative phosphorylation or more accurately, respiratory-chain phosphorylation. There are three sites within the mitochondrial chain which span the interaction between two neighbouring complexes, which on the basis of energy calculations are thought to witness a release of energy sufficient to synthesise almost one molecule of ATP from ADP and phosphate, as a result of electron transfer from one complex to its neighbour. These are designated site I between NADH and coenzyme Q, site II between *cytochromes b* and *c* and site III between cytochrome a and free oxygen. Site three occurs within complex IV, the final complex which may also be referred to as *cytochrome oxidase*. Its overall function is to transfer electrons from cytochrome c to cytochrome a, then to a_3 and finally to molecular oxygen. It is this final stage which is blocked by the action of cyanide and by carbon monoxide. Associated with the electron flow, is the ejection of hydrogen ions from inside the mitochondrion, across the membrane, and in complex IV, the reduction of the oxygen molecule with two hydrogen ions originating from inside the mitochondrion.

If all three sites were involved, the amount of energy released is sufficient to drive the synthesis of two and a half molecules of ATP for each pair of electrons transported. If the first site was omitted, the number falls to one and a half. In neither case is it a complete integer because there is not a direct mole for mole relationship between electron transport and ATP synthesis but as described earlier, it is part of a much more complicated process described above as the chemiosmotic theory.

Bacterial electron transport systems and oxidative phosphorylation

Bacterial electron transport chains have fundamentally the same function as that described for mitochondrial electron transport chains but with several notable differences in their structure. For example, the cytochrome oxidase, which is the final complex nearest the oxygen in mitochondria, is not present in all bacteria. The presence or absence of this complex is the basis of the 'oxidase' test for the identification of bacteria. In these organisms, cytochrome oxidase is replaced by a different set of cytochromes. An interesting example is *Escherichia coli*, an enteric bacterium and coliform, which is commonly found in sewage. It has replaced the electron carriers of cytochrome oxidase with a different set including cytochromes b_{558}, b_{595}, b_{562}, d and o, which are organised in response to the level of oxygen in the local environment. Unlike the mitochondrial chain, the bacterial systems may be highly branched and may have many more points for the entry of electrons into the chain and exit of electrons to the final electron acceptor.

Bacterial electron transport systems, denitrification and methanogenesis

As previously mentioned, the term respiration is applied to many processes. Without further specification it is usually used to mean the consumption of molecular oxygen, by reduction to water in the case of the electron transport discussed above, or by oxidation of an organic molecule to produce carbon dioxide and serine in the case of photorespiration, discussed later in this chapter. Thus the term anaerobic respiration seems a contradiction. It does however describe fundamentally the same process of electron transfer to a final acceptor which although inorganic, in this case is not oxygen. An example of such an electron acceptor is nitrate which is converted to nitrite. This is a toxic substance, and so many bacteria have the facility to convert nitrite to nitrogen gas. This overall series of reactions is described as denitrification and is the basis of the process by which denitrifying bacteria such as members of the *Pseudomonas* and *Bacillus* genera are able to reduce nitrate and nitrite levels down to consent values during sewage treatment. Such bacteria have different components in their electron transport chain in comparison with mitochondria, which have the necessary enzymatic activities to carry out these processes. Like mitochondrial electron transport, denitrification can be associated with synthesis of ATP although with much reduced efficiency.

Other examples of terminal electron acceptors are firstly sulphate, in which case one of the final products is elemental sulphur. This process is carried out by the obligate anaerobe, *Desulfovibrio* and members of the archaean genus *Archaeglobus*. Another anaerobe, *Alkaliphilus transvaalensis*, an extreme alkaliphile, growing at a pH of 8.5–12.5, isolated from an ultra-deep gold mine in South Africa, can use elemental sulphur, thiosulphate or fumarate as an additional electron acceptor (Takai *et al.*, 2001). Secondly, carbon dioxide may be the final electron acceptor in which case one of the final products is methane. This process is also carried out by obligate anaerobes, in this case, the methanogens, all of which are archaeans and are responsible for methane production in anaerobic digesters and landfill sites. Again, it functions on much the same principles as the other chains mentioned above but has a different set of cofactors which are most unusual. For both of the above obligate anaerobes, anaerobic respiration is an important mechanism of ATP synthesis. It is less efficient than aerobic respiration due to the smaller drop in electro-potential between sulphate or carbon dioxide and NADH compared with the difference between NADH and oxygen, and so less energy is available to be released during electron transport and consequently less ATP is synthesised per mole of NADH entering the pathway. Anaerobic respiration is however more efficient than fermentation and so is the route of choice for ATP synthesis for an anaerobe.

The energy balance sheet between substrate level and electron transport linked ATP synthesis

An approximate comparison may be made between the efficiency with respect to energy production, of ATP synthesis by substrate level phosphorylation and by association with electron transport. For 1 mol of glucose passing through glycolysis by the Embden-Meyerhof pathway to produce 2 mol of pyruvate, there is net production of 2 mol of ATP. For most fermentation pathways, no further ATP is synthesised. There are exceptions, of course, such as the conversion of an acyl CoA derivative such as acetyl CoA or butyryl CoA to the free acid which in these cases are acetate and butyrate respectively. Each of these reactions releases sufficient energy to drive the phosphorylation of 1 mol of ADP. Conversely, if the electron transport chain is functioning, NADH may be oxidised by relinquishing electrons to the cytochromes in the chain thus regenerating the oxidised cofactor. In this scenario, pyruvate may enter the TCA cycle rather than a fermentation route, thus a further mol of ATP is produced at substrate level during conversion of succinyl CoA to succinate via GTP, which then transfers the terminal phosphate to ATP. In addition, NADH and $FADH_2$ produced during the TCA cycle thus generating up to 15 mol of ATP per mol of pyruvate. An overall comparison may be made between glycolysis followed by reoxidation of NADH by fermentation or, alternatively, glycolysis followed by entry into the TCA cycle and reoxidation of cofactors via the electron transport chain. Remembering that 1 mol of glucose generates 2 mol of pyruvate during glycolysis, and that the 2 mol of NADH produced during glycolysis may also be reoxidised by transfer to the electron transport chain and not through fermentation, the net result is that glucose catabolised by the glycolysis-fermentation route results in the production of 2 mol of ATP whereas catabolism by the glycolysis-TCA cycle-electron transport/oxidative phosphorylation route produces up to 32 mol of ATP. The figure of 36 was deduced by Lehninger (1975) but has since been revised to reflect the tenets of the chemiosmotic theory described earlier.

Anaerobic respiration is less efficient than aerobic respiration. Oxidation of the same amount of cofactor by methanogenesis rather than oxidative phosphorylation would produce fewer mol of ATP. Consequently, for a given amount of ATP production, the flux of glucose through glycolysis followed by fermentation would have to be approximately 16 times greater than through glycolysis followed by oxidative phosphorylation, and the flux through methanogenesis is somewhat intermediate. It is the metabolic capability of the organism and the presence or absence of the appropriate inorganic electron acceptor which determines the fate of pyruvate on the grounds of energy considerations. On a practical basis this may explain why anaerobic processes, such as the anaerobic digestion of sewage sludge and municipal solid waste, are considerably less exothermic than their aerobic counterparts. For a given quantity of carbon source, an aerobic process will be able to extract in the order of 10 times the amount of energy than that generated by an anaerobic process.

Regeneration of NAD^+ in plants

In addition to the processes discussed above for the production of NADH, plant mitochondria operate an additional system whereby the required protons are derived from two molecules of the amino acid glycine. During this mitochondrial process, one molecule of molecular oxygen is consumed in the production of carbon dioxide and the amino acid, serine. The superfluous amino group from the second glycine molecule is released as ammonia. The glycine molecules were derived from phosphoglycolate, the metabolically useless product of photorespiration. This subject is very important with regard to plant breeding and development and so is discussed in some detail alongside the related subject of photosynthesis.

Photosynthesis and the Basis of Phytotechnology

The sun is the biosphere's ultimate source of energy and photosynthesis is the only means there is on this planet to trap incident sunlight and convert it into chemical energy available to biological processes. Thus, with very rare exceptions, organisms which do not photosynthesise, which is the majority, are totally dependent on those which do. With this introduction it is hardly surprising to find a description of this process in a book which specifically addresses the capabilities of biological organisms and their interplay. Leafy plants obviously feature in this section but so too do photosynthetic eukaryotic microorganisms and bacteria. A knowledge of this vital process is essential to appreciating the role which photosynthesising organisms play in the environment, their limitations and the strengths upon which biotechnology can capitalise.

This process is used to drive all the biochemical synthesis and degradation reactions occurring in the cell in addition to various other energy requiring processes such as the movement and transport of molecules across membranes. Energy is finally dissipated as heat, and entropy rises in accordance with the Laws of Thermodynamics. Any interference with the flow from the sun either by reducing the ability of the energy to penetrate the atmosphere, or by reducing the total photosynthetic capacity of the planet, has dramatic consequences to all forms of life. Conversely, too intense a radiation from the sun resulting from thinning of the ozone layer runs the risk of damaging the photosynthetic machinery. This can be compensated for by the organism acquiring pigments to absorb harmful radiation, but it requires time for such an evolutionary adjustment to take place.

It is noteworthy that the bulk of photosynthesis is performed by unicellular organisms, such as photosynthetic algae, rather than the macrophytes as might reasonably be supposed. Photosynthesis occurs in two parts; the first is the trapping of light with associated reduction of $NADP^+$ and ATP synthesis, and the second is the fixing of carbon dioxide by its incorporation into a carbohydrate molecule. This is most commonly a hexose sugar, and typically glucose, the synthesis of which utilises the NADPH and ATP produced in the light dependent part 1. The processes of carbohydrate synthesis occurring in the second part are described as the dark reactions, so called because they may proceed in the dark after a period of illumination to activate part 1. The sugar produced during these dark reactions will then be utilised by the cell, transferred to another cell or ingested by a larger organism and eventually catabolised to carbon dioxide and water releasing the energy consumed originally to synthesise the molecule. Here is another example of a natural cycle, where carbon is introduced, as carbon dioxide, into the synthesis of a sugar which is then interconverted through the various metabolic pathways until finally it is released as carbon dioxide thus completing the cycle. Eukaryotes capable of carrying out photosynthesis include higher green plants, multicellular green, brown and red algae and various unicellular organisms such as the euglenoids and dinoflagellates, both of which are commonly found in fresh water environments, and diatoms which are also found in salt water. The diatoms which are unicellular algae, are particularly noteworthy given estimates that they are responsible for fixing 20–25% of the world's carbon through photosynthesis (Round, Crawford and Mann, 1990). Prokaryotes capable of photosynthesis include blue-green algae, and both the sulphur and non-sulphur purple and green bacteria. The blue-green algae which are oxygenic bacteria and are alternatively named cyanobacteria, operate light reactions very similar to those of eukaryotes. Conversely, the green and the purple non-sulphur bacteria which are both facultative aerobes and the strictly anaerobic green and the purple sulphur bacteria utilise a rather different set of light reactions as a consequence of their possessing a 'simpler' photosystem. Eukaryotic and bacterial systems are both described in the following sections.

The light reactions

Visible light is the outcome of the nuclear fusion of hydrogen atoms, resulting in the production of helium atoms, gamma radiation and two electrons. This fusion occurs in the sun at a temperature of approximately 20 000 000 K. The gamma radiation and electrons combine to produce quanta of visible light. The entrapment of light is performed in photosynthetic cells by pigments; the most important of which are the chlorophylls. These are flat ring structures, with regions of conjugated double and single bonds, and a long hydrophobic tail well designed for anchoring the pigments into membrane. Only red and blue light is absorbed by the chlorophylls in most organisms, consequently, when white light from the sun shines upon them, they reflect green light, thus making these organisms appear green. Variation in the types of chlorophylls and the presence of additional accessory pigments all contribute to the observed colour of the organism and are the result of evolution which has developed the 'best fit' of light trapping molecules to suit the ecological niche of the organism. It is worth pointing out that wholesale transport of the plant or bacterium for biotechnology purposes has to take this factor into account. It is important to test the growth and performance characteristics of any translocated plant or bacterium to ensure that the new environment does not produce disappointing results. This problem is addressed with respect to choice of *Phragmites* species in the case study on reed beds in Chapter 7. The purpose of the accessory pigments referred to above, which include the carotenoids and phycobiliproteins, the latter found in red algae and cyanobacteria, is to extend the range of absorbed wavelengths thus maximising the amount of energy trapped from light and protecting the photosynthetic system from potential damage by oxidation. A rather unusual pigment which functions as a primary pigment, is bacteriorhodopsin which causes the archaea which express it, to appear purple. Returning to the principally eukaryotic process, the chlorophylls described above which receive the incident light are clustered in highly organised structures called antennae located on the cell surface. The incident light excites the energy state of the recipient chlorophyll to a higher energy state. When the chlorophyll returns to its normal level, it releases electrons which are transferred to a neighbouring chlorophyll. The transfer is repeated until the electrons arrive at a photosystem, from which point they enter an electron transport chain linked to the reduction of NAD$^+$ and the synthesis of ATP. There are many similarities between electron transport in respiration and in photosynthesis in that they are both membrane bound and may be coupled to phosphorylation and thus synthesis of ATP, according to Mitchell's chemiosmotic theory, by employing a similar strategy of a proton gradient described earlier for respiration. In eukaryotic higher organisms, photosynthesis occurs in the chloroplast while in bacteria it occurs in the cytoplasmic membrane. The precise location in bacteria is sometimes described as being the mesosome. This has been reported as an infolding of the bacterial cellular membrane which sometimes appears to be in association with the bacterial DNA and often is found near nascent cell walls. Although considerable effort has been invested in determining its function, there is still disagreement as to whether or not the mesosome is indeed a bacterial cell structure or is simply an artefact occurring during preparation of samples for microscopy (Fernandez *et al.*, 2005). Thus the site of bacterial photosynthesis remains uncertain beyond it being bound in the cytoplasmic membrane.

Photosystems in eukaryotes and cyanobacteria

There are two types of photosystem which may occur in photosynthetic organisms indicated in Figure 2.10; photosystem 1 which receives electrons from photosystem 2 but may also operate independently by cyclic electron transport, and photosystem 2 which is not present in all such organisms. The pathway for electron transport has two principal routes. One involves only photosystem 1. In this, electrons transferred from the antennae to photosystem 1 cause excitation of the chlorophyll in this system. When the chlorophyll returns to its lower energy state, the electrons are transferred to ferredoxin, which is one of the iron containing proteins in the chain of electron carriers. From this point there is a choice of routes; either the cyclic path by way of a chain of cytochrome molecules starting with cytochrome b$_{563}$ and finally returning to chlorophyll a, or by the non-cyclic route which is the transfer of electrons to NADP$^+$.

The source of the hydrogen atom required to reduce the NADP$^+$ to NADPH in this system is the water molecule which donates its electrons to photosystem 2 to replace those lost to NADP$^+$. It is the origin of the oxygen released during photosynthesis, hence the term oxygenic. Thus the overall flow of electrons in the non-cyclic pathway is from the water molecule, through photosystem 2, along a series of cytochromes to photosystem 1 and thence to ferredoxin and

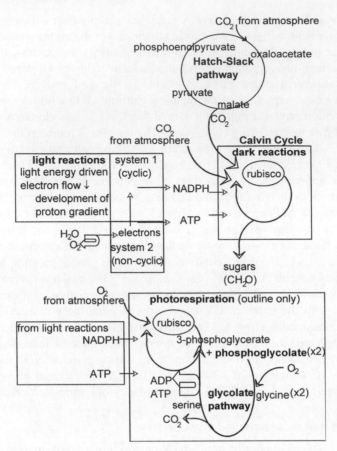

Figure 2.10   *Oxygenic photosynthesis, photorespiration and the Hatch–Slack pathway*

finally NADP$^+$, which also collects a hydrogen atom to complete the reduction to NADPH. Both the cyclic and the non-cyclic pathways produce a proton gradient which drives the synthesis of ATP, but only the cyclic route has the facility of producing NADPH. A combination of cyclic and non-cyclic pathways are used by the organism to produce the required amounts of NADPH and ATP used in the dark reactions for the synthesis of carbohydrate. So far the description has been of photosynthesis in eukaryotes and cyanobacteria.

Photosystems in purple and green bacteria

Looking at the general equation for the chemical reactions in photosynthesis shown in Figure 2.11, it may be seen that water is the electron donor during oxygenic photosynthesis while several molecules may play the same role in anoxygenic systems. Suitable molecules are listed in the figure, from which several interesting observations may be made. If the electron donor is

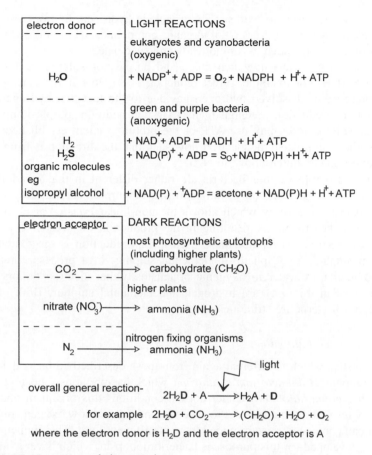

Figure 2.11 *Reactions of photosynthesis*

hydrogen sulphide, which is the principle gas responsible for the foul smell, reminiscent of rotten eggs, typically found in wet and untilled soil, for example in the bottom of ponds, the product is sulphate or elemental sulphur. Examination of bacterial photosystems explains how this occurs.

Green and purple bacteria possess only one photosystem which is a fairly basic equivalent to photosystem 1 of eukaryotes and cyanobacteria, but employing a different set of electron carriers. In purple non-sulphur bacteria, this is only capable of cyclic electron flow which produces a proton gradient and thus allows ATP photosynthesis, but this process does not lead to NADPH production. A similar system exists in green non-sulphur bacteria. The lack of a photosystem equivalent to photosystem 2 in eukaryotes, requires these bacteria to provide a different route for the regeneration of NADH which serves much the same function as NADPH in carbohydrate synthesis. Their solution to this problem is to use as electron donors, molecules which have a more negative reduction potential than water and so are easier to oxidise. These include hydrogen, hydrogen sulphide, elemental sulphur and a variety of organic compounds including sugars, and various organic acids such as amino acids and succinate.

There are many ways in which green and purple non-sulphur bacteria may produce NADH. For example, direct reduction is possible if they are growing in the presence of dissolved hydrogen gas due to the fact that hydrogen has a more negative reduction potential than NAD^+. In addition, purple non-sulphur photosynthetic bacteria may use ATP or the proton gradient established during photosynthesis to reverse the electron flow such that the direction is from one of the electron donors noted above to NAD^+.

Green and purple sulphur bacteria are rather different in that in addition to having a cyclic system broadly similar to that of purple bacteria, they have an additional enzyme activity which allows the non-linear transfer of electrons to ferredoxin linked to NAD^+ resulting in the production of NADH. One of the sources of electrons to replace those used in this reduction is the oxidation of hydrogen sulphide to sulphate or elemental sulphur, in a process comparable to the oxidation of water in oxygenic organisms. Other electron donors which may be used in this way are hydrogen and elemental sulphur. Both of these non-sulphur bacteria are strict anaerobes.

Photosystem in a halophile

A photosystem which is different again from those described so far is that found in the halophile *Halobacterium salinarium* which has previously been classified as *Halobacterium halobium*. Under normal conditions this organism obtains its energy by respiration, but in order to survive conditions of low oxygen concentrations, it can photosynthesise provided there is sufficient light. The pigment which has been developed for this purpose is bacteriorhodopsin, which is very similar to the rhodopsin pigment found in vertebrate eyes. The part of the molecule which absorbs light is retinal. When this occurs, changes in the bond formation of this chromophore result in expulsion of protons across the membrane thus producing a proton gradient. As described for other systems, this proton gradient may then be used to drive ATP synthesis.

The dark reactions

The result of illumination of a photosynthetic organism is to stimulate electron transport which leads to the production of NADPH or NADH, and synthesis of ATP. Both are required for the next stage which in eukaryotes and cyanobacteria (blue-green algae) is the synthesis of sugar from carbon dioxide involving the Calvin Cycle. Many biochemistry textbooks give excellent descriptions of this process and so only a summary is given in Figure 2.10.

In brief, ribulose diphosphate is carboxylated with carbon dioxide catalysed by the enzyme *rubisco* to form an unstable six carbon sugar which is then cleaved to form two molecules of 3-phosphoglycerate, an intermediate of glycolysis. This is not the only route of entry of carbon dioxide into carbohydrate synthesis, the other being the Hatch Slack Pathway. This subject is discussed in more detail later in this chapter.

Returning to the Calvin Cycle, rearrangements of 3-phosphoglycerate produced by *rubisco* then take place by similar steps to the reversible steps of glycolysis and reactions of the pentose phosphate pathway. After completion of three cycles, the net result is the fixation of three molecules of carbon dioxide into a three carbon sugar, each cycle regenerating a ribulose phosphate molecule. After phosphorylation of the trioses at the expense of ATP, they may enter into glycolysis and be converted into glucose and then to starch to be stored until required. The Calvin Cycle is so familiar that it is easy to overlook the fact that not all reducing equivalents are channelled through rubisco to this cycle and carbohydrate synthesis.

Some organisms may use additional pathways involving other electron acceptors such as nitrate, nitrogen and hydrogen atoms, the reduction of which obviously does not produce carbohydrate but different essential nutrients which may then also be available to other organisms. These products are summarised in Figure 2.11. For example, when nitrogen or nitrate is the electron donor, the product is ammonia which becomes incorporated by amino transfer into amino acids and thus forms part of the nitrogen cycle. Nitrogen is a particularly noteworthy case in the context of this book as it is the process of nitrogen fixation. This is performed by a number of nitrogen fixing bacteria, some of which are free living in the soil and some form symbiotic relationships with certain leguminous plants, forming root nodules. Nitrogen fixation is, by necessity, an anaerobic process, and so one essential role for the plant is to provide a suitable oxygen-free environment for these bacteria, the other, to provide energy. Genetic manipulation of plants is discussed in Chapter 9 but it is relevant to point out here that the suggestion is often mooted of introducing the genes responsible for allowing nitrogen fixation to be transferred from the relevant bacteria into suitable plants. The symbiotic relationship between plant and bacteria is very difficult to create artificially and has been a stumbling block in the drive to increase the number of plant species able to host nitrogen fixation. The complicated interaction between plant and bacterium involves intricacies of plant physiology as well as genetic capability provided by the bacterium, and so it is unlikely that a simple transfer of nitrogen fixation genes from bacterium to plant will be successful. This, however, remains a research area of major importance.

The issue of nitrogenous material, particularly in respect of sewage and associated effluents, is of considerable relevance to the environmental application of biotechnology. In addition, there is great potential for phytotechnological intervention to control nitrogen migration, most especially in the light of the burgeoning expansion of nitrate sensitive areas within the context of agricultural fertiliser usage. Hence, bio-engineering of the nitrogen cycle, at least at the local level, provides an important avenue for the control of pollution and the mitigation of possible eutrophication of aquatic environments. The cycle itself, and some of the implications arising, are discussed later in this chapter.

C_3 and C_4 plants

Plants for which the reaction catalysed by *rubisco* is the first point of entry of atmospheric carbon dioxide into carbohydrate metabolism are termed C_3 plants due to the product of rubisco being two molecules of 3-phosphoglycerate which contains three carbons. This is the typical route for temperate organisms. An alternative to direct carboxylation for introducing carbon dioxide into the Calvin Cycle used by some tropical plants, is the Hatch-Slack Pathway illustrated in Figure 2.10. In this case the first step of entry for atmospheric carbon dioxide is by carboxylation of phosphoenolpyruvate by *phosphoenolpyruvate carboxylase* to produce the four carbon molecule, oxaloacetate. Hence, plants able to use this pathway are termed C_4 plants. The oxaloacetate is part of a cycle which carries the carbon dioxide into the bundle-sheath cells and so away from the surface of the plant, to where the oxygen concentration is lower. Here the carbon dioxide, now being carried as part of malate is transferred to *rubisco* thus releasing pyruvate which returns to the mesophyll cells at the surface of the plant where it is phosphorylated at the expense of ATP to phosphoenolpyruvate ready to receive the next incoming carbon dioxide molecule from the atmosphere.

The overall effect is to fix atmospheric carbon dioxide, transfer it to a site of lower oxygen concentration compared with the surface of the plant, concentrate it in the form of malate and then transfer the same molecule to *rubisco* where it enters the Calvin Cycle. Although the Hatch Slack Pathway uses energy, and therefore may seem wasteful, it is of great benefit to plants growing in the warmer regions of the globe. The reason for this is that the enzyme involved in carbon dioxide fixation in C_4 plants namely, *phosphoenolpyruvate carboxylase* has a very high affinity for carbon dioxide and does not use oxygen as a substrate, contrasting with *rubisco*. The result of this competition between oxygen and carbon dioxide for binding to *rubisco* is the futile process of photorespiration, described in the next section. The affinity of carbon dioxide for *rubisco* falls off with increasing temperature and so in a tropical environment, the efficiency of *rubisco* to fix carbon dioxide is low. In this situation, the disadvantage in using energy to operate the Hatch Slack Pathway is more than compensated for by the advantage of being able to fix carbon dioxide efficiently at elevated temperatures. So advantageous is this that much research is being directed to transferring the capability to operate the Hatch Slack Pathway into selected C_3 plants.

In the broadest sense of environmental biotechnology, the potential maximisation of solar energy usage, either as a means to the remediation of contamination or to reduce potential pollution by, for example excessive fertiliser demand, could be of considerable advantage. Hence, appropriately engineered C_3 plants in either role offer major advantages in solar efficiency, which, in temperate climes, could provide significant environmental benefits.

Photorespiration

Returning to synthesis of carbohydrate by the Calvin Cycle, as mentioned above, the first step is the carboxylation of the five carbon sugar, ribulose diphosphate catalysed by rubisco. As mentioned in the preceding section, this enzyme may also function as an oxidase indicated by its full name ribulose diphosphate carboxylase oxidase. When this occurs and oxygen replaces carbon dioxide, the ensuing reaction produces phosphoglycolate in addition to 3 phosphoglycerate. Since, as a result of illumination, oxygen is consumed and carbon dioxide is released during the reactions of the glycolate pathway, this process is termed photorespiration and occurs alongside photosynthesis. The higher the ambient temperature in which the organism is growing, and the higher the oxygen concentration relative to carbon dioxide, the more pronounced the oxidase activity becomes and consequently the less efficient rubisco is at introducing carbon dioxide into carbohydrate synthesis. The phosphoglycolate formed as a result of oxygen acting as substrate for rubisco, is then dephosphorylated to form glycolic acid. There follows a series of reactions forming a salvage pathway for the carbons of glycolic acid involving transfer of the carbon skeleton to the peroxisomes, then to the mitochondria, back to the peroxisomes and finally back to the chloroplast in the form of glycerate which is then phosphorylated at the expense of ATP to re-enter the Calvin Cycle as 3-phosphoglycerate. The result of this detour, thanks to the oxidase activity of rubisco is to lose a high energy bond in phosphoglycolate, to consume ATP during phosphorylation to produce 3-phosphoglycerate, consumption of oxygen and release of carbon dioxide. This pathway resulting from the oxidase activity of rubisco, shown in Figure 2.10 is wasteful because it consumes energy obtained by the light reactions with no concomitant fixation of carbon dioxide into carbohydrate. The C_3 plants are therefore operating photosynthesis under suboptimal conditions especially when the oxygen tension is high and carbon dioxide tension is low. Why rubisco has not evolved to lose the oxidase activity is unclear: presumably evolutionary pressures of competition have been insufficient to date. For the reasons indicated in the preceding section, C_4 plants show little or no photorespiration due to their ability to channel carbon dioxide to rubisco by a method independent of oxygen tension. Therefore they are considerably more efficient than their C_3 counterparts and may operate photosynthesis at much lower concentrations of carbon dioxide and higher concentrations of oxygen. It is interesting to contemplate the competitive effects of introducing C_4 style efficiency into C_3 plants, but at the moment this is just speculation.

Balancing the light and the dark reactions in eukaryotes and cyanobacteria

Using the 6 carbon sugar, glucose, as an example, synthesis of 1 molecule requires 6 carbon dioxide molecules, 12 molecules of water, 12 protons, 18 molecules of ATP and 12 molecules of NADPH. Since photophosphorylation is driven by a proton gradient established during electron flow after illumination, there is not a stoichiometric relationship between the number of photons exciting the systems and the amount of ATP produced. However it is now established that for every eight photons incident on the two photosystems, four for each system, one molecule of oxygen is released, two molecules of $NADP^+$ are reduced to

NADPH and approximately three molecules of ATP are synthesised. Since this may leave the dark reactions slightly short of ATP for carbohydrate synthesis, it is postulated that photosystem 2 passes through one extra cycle thus producing additional ATP molecules with no additional NADPH.

The Nitrogen Cycle

Nitrogen is constantly taken, or fixed, from the atmosphere, oxidised to a form able to be utilised by plants and some bacteria, to be subsumed into metabolic pathways, and through the various routes described above is then excreted into the environment as reduced nitrogen where it may be reoxidised by bacteria or released back into the atmosphere as nitrogen gas. These combined processes are known collectively as the nitrogen cycle.

The previous discussions have referred to the release of nitrogen during degradation of proteins and nucleic acid bases, either in the form of ammonia, the ammonium ion, urea or uric acid. The fate of all these nitrogen species is to be oxidised to nitrite ion by Nitrosomas, a family of nitrifying bacteria. The nitrite ion may be reduced and released as atmospheric nitrogen, or further oxidised to nitrate by a different group of nitrifying bacteria, Nitrobacter. The process of conversion from ammonia to nitrate is sometimes found as a tertiary treatment in sewage works to enable the nitrate consent to be reached. The process typically occurs in trickling bed filters which have, over time, become populated with a Nitrosomas and Nitrobacter along with the usual flora and fauna which balance this ecosystem. Denitrification may then occur to release atmospheric nitrogen or the nitrate ion, released by Nitrobacter, may be taken up by plants or some species of anaerobic bacteria where it is reduced to ammonium ion and incorporated into amino acids and other nitrogen-carbon containing compounds. To complete the cycle, atmospheric nitrogen is then fixed by nitrifying bacteria, either free living in the soil or in close harmony with plants as described earlier in this chapter.

Closing Remarks

The underpinning biochemistry and natural cycles described in this chapter form the basis of all environmental biotechnological interventions, and a thorough appreciation of them is an essential part of understanding the practical applications which make up most of the rest of this work.

References

Allison, D.G., Gilbert, P., Lappin-Scott, H. and Wilson, M. (2000) Community structure and co-operation in biofilms, *Fifty-Ninth Symposium of the Society for General Microbiology held at the University of Exeter*, Cambridge University Press, Cambridge, pp. 215–256, September 2000.

Barrow, G.I. and Feltham, R.K.A. (1993) *Cowan and Steel's Manual for the Identification of Medical Bacteria*, 3rd edn, Cambridge University Press, Cambridge.

Cavalier-Smith, T. (2002) The neomuran origin of archaebacteria, the negibacterial root of the universal tree and bacterial megaclassification. *International Journal of Systematic and Evolutionary Microbiology*, **52**, 7–76.

Demaneche, S., Bertolla, F., Buret, F. *et al.* (2001) Laboratory-scale evidence for lightning-mediated gene transfer in soil. *Applied and Environmental Microbiology*, **67**, 3440–3444.

Ehlers, L.J. (2000) Gene transfer in biofilms community structure and co-operation in biofilms, *Fifty-Ninth Symposium of the Society for General Microbiology held at the University of Exeter*, Cambridge University Press, Cambridge, pp. 215–256, September 2000.

Fernandez, M.I., Prevost, M-C., Sansonetti, P.J. and Griffiths, G. (2005) Applications of cryo- and transmission electron microscopy in the study of microbial macromolecular structure and bacterial-host cell interactions, *Methods in Microbiology*, Vol. 34, Elsevier B V, Amsterdam, pp. 137–162.

Horinouchi, M., Yamamoto, T., Taguchi, K. *et al.* (2001) *Meta*-cleavage enzyme gene *tes*B is necessary for testosterone degradation in *Comamonas testosteroni* TA441. *Microbiology*, **147**, 3367–3375.

Köhler, R.H., Cao, J., Zipfel, W.R. *et al.* (1997) Exchange of protein molecules through connections between higher plant plastids. *Science*, **276**, 2039–2042.

Kreft, J.-U., Picioreanu, C., Wimpenny, J.W.T. and van Loosdrecht, M.C.M. (2001) Individual-based modelling of biofilms. *Microbiology*, **147**, 2897–2912.

Lehninger, A.L. (1975) *Biochemistry*, 2nd edn, Worth, New York, pp. 536–538.

Mandelstam, J. and McQuillen, K. (1973) *Biochemistry of Bacterial Growth*, 2nd edn, Blackwell Scientific Publications, Oxford, p. 166.

McMaster, M.E. (2001) A review of the evidence for endocrine disruption in Canadian aquatic ecosystems. *Water Quality Research Journal of Canada*, **36**, 215–231.

Michal, G. (1992) *Biochemical Pathways*, 3rd edn, Boehringer Mannheim GmbH.

Round, F.E., Crawford, R.M. and Mann, D.G. (1990) *The Diatoms*, Cambridge University Press, Cambridge.

Stackebrandt, E. and Woese, C.R. (1981) The evolution of prokaryotes, molecular and cellular aspects of microbial evolution, *Thirty-Second Symposium of the Society for General Microbiology held at the University of Edinburgh 1981*, Cambridge University Press, Cambridge, pp. 1–31.

Takai, K., Moser, D.P., Onstott, T.C. *et al.* (2001) *Alkaliphilus transvaalensis* gen.nov., sp. nov., an extremely alkaliphilic bacterium isolated from a deep South African gold mine. *International Journal of Systematic and Evolutionary Microbiology*, **51**, 1245–1256.

Whittam, T.S. (1992) Sex in the soil. *Current Biology*, **2**, 676–768.

Woese, C.R. and Fox, G.E. (1977) Phylogenetic structure of the prokaryotic domain: the primary kingdoms. *Proceedings of the National Academy of Sciences of the United States America*, **74**, 5088–5090.

Woese, C.R., Kandler, O. and Wheelis, M.L. (1990) Towards a natural system of organisms: proposal for the domains Archaea, Bacteria, and Eucarya. *Proceedings of the National Academy of Sciences of the United States America*, **87**, 4576–4579.

Case Study 2.1   Microbial Eco-Toxicology Testing (Scotland/Sweden)

Assessment of eco-toxicant effects has often been made conventionally by extrapolating from straightforward chemical analysis of the soil, water or other medium being investigated, but while this is useful information *per se* it gives little clue as to their impact on biological systems.

Although there is an obvious potential in extending microbial assays to the problem, many of the available techniques have made use of individual species, typically depending on a single genotypic character, such as the Lux operon in the bio-luminescent bacterium, Vibrio fischeri. As a consequence, though the tests developed are swift and well established, they are undoubtedly open to criticism for their narrow basis. However, with support from Scottish Enterprise and the UK's EUREKA scheme, together with Sweden's Karolinska Institute, the Aberdeen-based biotech company, NCIMB, has recently succeeded in expanding the concept to embrace a diverse, multi-species complement of microbes.

The Microbial Assay for toxic Risk Assessment (MARA) kit uses 11 genetically diverse microbe species (1 yeast and 10 bacteria from the alpha, beta and gamma proteobacteria) lyophilised in situ in a 96-well micro-titre plate. In use, the micro-organisms are first reconstituted at 30 °C for 4 hours before being inoculated with a range of sample dilutions and incubated, again at 30 °C, for 18 hours. At the end of this stage the plate is scanned and dedicated image analysing software used to quantify microbial growth or inhibition by measuring the level of reduction of a redox dye incorporated in the plate's substrate medium.

This enables a microbial toxic concentration (MTC) value to be calculated for each individual microbial species as well as a mean MTC for the entire 11 micro-organisms, yielding a 'toxic fingerprint' for the sample, which could prove a particularly useful indicator in routine monitoring applications. The system also has considerable investigative potential to help diagnose unidentified contaminants in the environment, by enabling a comparative dendogram to be produced to allow the unknown chemicals to be evaluated against known standards.

3

Fundamentals of Biological Intervention

The manipulation of natural cycles lies at the heart of much environmental biotechnology and engineering solutions to the kinds of problem for which this technology is appropriate, typically centres on adapting existing organisms and their inherent abilities. For the most part, the sorts of 'environmental' problems that mankind principally concerns itself about, are those which exist in the portion of the biosphere which most directly affects humanity itself. As a result, most of the organisms used share many of our own needs and the majority of the relevant cycles are ones which are, at least, largely familiar. While other aspects of biotechnology may demand techniques of molecular biology and genetic manipulation, as has been discussed, the applications of biological science, certainly to questions of pollution and waste, generally do not. Their position in respect of the third leg of the intervention tripod, shown in Figure 1.1, in clean manufacturing, is more ambiguous and there is distinct scope for them to have a greater role here, in the future. However, while this undoubtedly represents a contribution in terms of reduced pollution or the minimisation of waste, with regard to the express demands for environmental amelioration, their involvement is, at best, marginal. This is not to say that genetically manipulated organisms (GMOs) have no relevance to the field, but rather that, on the whole, it is greatly eclipsed in much of current practice by rather more ordinary organisms.

Using Biological Systems

Consequently, a number of themes and similarities of approach exist, which run as common and repeated threads throughout the whole of the science. Thus, optimisation of the activities of particular organisms, or even whole biological communities, to bring about any desired given end, typically requires manipulation of local conditions. Control of temperature, the accessibility of nutrients and the availability of oxygen are commonly the tools employed, especially when the target effectors are microbes or isolated biological derivatives. For the kind of whole organism approaches typified by phytotechnological interventions

Environmental Biotechnology: Theory and Application, Second Edition Gareth M. Evans and Judith C. Furlong
© 2011 John Wiley & Sons, Ltd.

discussed in Chapter 7, this may prove a more difficult proposition, but never-the-less, one which still remains relevant at least in principle. The typical factors affecting the use of biological systems in environmental engineering relate to the nature of the substances needing to be removed or treated and to the localised environmental conditions pertaining to the particular situation itself. Thus, in respect of the former, the intended target of the bio-processing must generally be both susceptible and available to biological attack, in aqueous solution, or at least in contact with water, and within a low to medium toxicity range. Generally, the local environmental conditions required would ideally offer a temperature of 20–30 °C but a range of 0–50 °C will be tolerated in most cases, while an opti-mum pH lies in the range 6.5–7.5, but again a wider tolerance of 5.0–9.0 may be acceptable, dependent on the precise organism involved. For land based appli-cations, especially in the remediation of contamination or as a component of integrated pollution control measures, there is an additional common constraint on the substrate. Typically the soil types best suited to biotechnological inter-ventions are sands and gravels, with their characteristically low nutrient status, good drainage, permeability and aeration. By contrast, biological treatments are not best suited to use in clays or peaty or other soils of high organic content. In addition, generalised nutrient availability, oxygenation and the presence of other contaminants can all play a role in determining the suitability of biological intervention for any given application.

Extremophiles

As has been previously mentioned, in general the use of biotechnology for envi-ronmental management relies on mesophilic micro-organisms which have roughly similar environmental requirements to ourselves, in terms of temperature, pres-sure, water requirement and relative oxygenation. However, often some of their abilities, which are directly instrumental in enabling their use in this context, arose in the first instance as a result of previous environmental pressures in the species (pre)history. Accordingly, ancient metabolic pathways can be very valuable tools for environmental biotechnology. Thus, the selective advantages honed in Carboniferous coal measures and the Pleistocene tar pits have produced microbes which can treat spilled mineral oil products in the present and methano-genesis, a process developed by the Archae during the dawn of life on Earth, remains relevant to currently commonplace biological interventions. Moreover, some species living today tolerate extreme environments, like high salinity, pres-sures and temperatures, which might be of use for biotech applications requiring tolerance to these conditions. The Archaea (the group formerly known as the archaebacteria and now recognised as forming a distinct evolutionary line) rank amongst their numbers extreme thermophiles and extreme halophiles in addition to the methanogens previously mentioned. Other species tolerate high levels of ionising radiation, pH or high pressures as found in the deep ocean volcanic vents known as 'black smokers'.

Making use of these extremophile organisms could provide a way of developing alternative routes to many conventional chemicals or materials in such a way as to offer significant advantages over existing traditional processes. Many current industrial procedures generate pollution in one form or another and the challenge of such 'green chemistry' is to design production systems which avoid the potential for environmental contamination. The implementation of 'clean manufacturing technologies' demands considerable understanding, innovation and effort if biologically derived process engineering of this kind is to be made a reality. With environmental concerns placing ever growing emphasis on energy efficiency and low carbon usage, industrial applications of the life sciences in this way seem likely to be increasingly relevant. To date, however, there has been little commercial interest in the extremophiles, despite their very obvious potential for exploitation.

The existence of microbes capable of surviving in extreme environments has been known since the 1960s, but the hunt for them has taken on added impetus in recent years as possible industrial applications for their unique biological capabilities have been recognised. As might be expected, much of the interest centres on the extremophile enzymes, the so-called 'extremozymes', which enable these species to function in their demanding natural habitats. The global market for enzymes amounts to around $3 billion (US) annually for biomedical and other industrial uses and yet the 'standard' enzymes typically employed cease working when exposed to heat or other extreme conditions. This often forces manufacturing processes that rely on them to introduce special steps to protect the proteins during either the active stage or storage. The promise of extremozymes lies in their ability to remain functional when other enzymes cannot. The potential for the mass use of enzymatic 'clean production' is discussed more fully in the following chapter, but the major benefit of using extremophile enzymes in this role is that they offer a way to obviate the requirement for such additional procedures, which inevitably both increases process efficiency and reduces costs. In addition, their novel and distinct abilities in challenging environments allows them to be considered for use as the basis of entirely new enzyme-based approaches to processing. Such methods, if properly designed and implemented, have the potential to give rise to major environmental and economic benefits compared with traditional energy-intensive chemical procedures. However, the widespread uptake and integration of biocatalytic systems as industrial production processes in their own right is not without obstacles which need to be overcome. In many conventional catalytic processes, chemical engineers are free to manipulate turbulence, pH, temperature and pressure for process intensification, often using a variety of reactor configurations and regimes to bring about the desired enhancement of productivity (Wright and Raper, 1996). By contrast, in biological systems, the use of turbulence and other such conventional intensification methods are not appropriate as the microbial cells are typically too sensitive to be subjected to this treatment, as are the isolated enzymes. Such procedures often irreversibly denature proteins, destroying enzymatic activity.

Thermophiles

Of all the extremophiles, thermophiles are amongst the best studied, thriving in temperatures above 45 °C, with some of their number, termed hyperthermophiles, prefer temperatures in excess of 85 °C. Unsurprisingly, the majority of them have been isolated from environments which have some association with volcanic activity. The first extremophile capable of growth at temperatures greater than 70 °C was identified in the late 1960s as a result of a long-term study of life in the hot-springs of Yellowstone National Park, Wyoming USA, headed by Thomas Brock of the University of Wisconsin-Madison. Now known as *Thermus aquaticus*, this bacterium would later make possible the widespread use of a revolutionary technology, the polymerase chain reaction (PCR), which is returned to later in this chapter. Shortly after this initial discovery, the first true hyperthermophile was found, this time an archaean which was subsequently named *Sulfolobus acidocaldarius*. Having been discovered in a hot acidic spring, this microbe thrives in temperatures up to 85 °C. Hyperthermophiles have since been discovered from deep sea vent systems and related features such as geothermal fluids, attached sulphide structures and hot sediments. Around 50 species are presently known. Some grow and reproduce in conditions hotter than 100 °C, the current record being held by *Pyrolobus fumarii*, which was found growing in oceanic 'smokers'. Its optimum temperature for reproduction is around 105 °C but will continue to multiply up to 113 °C. It has been suggested that this represents merely the maximum currently accepted for an isolated and culturable hyperthermophile and is probably not even close to the upper temperature limit for life which has been postulated at around 150 °C, based on current understanding. Although no one knows for certain at this time, it is widely thought that higher than this the chemical integrity of essential molecules will be unlikely to escape being compromised.

To set this in context, isolated samples of common place proteins, like egg albumin, are irreversibly denatured well below 100 °C. The more familiar mesophilic bacteria enjoy optimum growth between 25 and 40 °C; no known multicellular organism can tolerate temperatures in excess of 50 °C and no eukaryotic microbe known can survive long term exposure to temperatures greater than around 60 °C. The potential for the industrial exploitation of the biochemical survival mechanisms which enable thermo- and hyperthermophiles to thrive under such hot conditions is clear. In this respect, the inactivation of thermophiles at temperatures which are still too hot for other organisms to tolerate may also have advantages in commercial processes. Though an extreme example in a world of extremes, the previously mentioned *P. fumarii*, stops growing below 90 °C; for many other species the cut-off comes at around 60 °C.

A good understanding of the way in which extremophile molecules are able to function under these conditions is essential for any future attempt at harnessing the extremozymes for industrial purposes. One area of interest, in particular is how the structure of molecules in these organisms, which often very closely resemble their counterparts in mesophilic microbes, influences activity. In a number of heat-tolerant extremozymes, for example the major difference appears to be no more than an increased prevalence of ionic bonds within the molecule.

Though the industrial use of extremophiles in general has been limited to date, it has notably given rise to PCR, a major technique used in virtually every molecular biology laboratory worldwide. The application of PCR has, in addition, opened the flood gates for the application of genetic analyses in many other branches of life science, including forensics and medical diagnosis. Though this is a tool of genetic engineering rather than anything which could be argued as an 'environmental' application, it does illustrate the enormous potential of extremozymes. The process uses a DNA polymerase, called Taq polymerase, derived from *T. aquaticus*, as mentioned earlier, and was invented by Kary Mullins in the mid 1980s. The original approach relied on mesophilic polymerases and since the reaction mixture is alternately cycled between low and high temperatures, enzymatic denaturation took place, requiring their replenishment at the end of each hot phase. Samples of *T. aquaticus* had been deposited shortly after the organism's discovery, some 20 years earlier, and the isolation of its highly heat tolerant polymerase enabled totally automated PCR technology to be developed. More recently, some PCR users have begun to substitute Pfu polymerase, isolated from another hyperthermophile, *Pyrococcus furiosus*, which has an optimum temperature of 100 °C. One area, however, where thermophiles could possibly come into their own in future is in the production of clean energy, either in terms of bioethanol production from hemicellulose or in continuous hydrogen production for conversion in a fuel cell. The former has been investigated using a number of anaerobic thermophiles, including new isolates of strains growing optimally at 70–80 °C, for their ethanol production from d-xylose (Sommer, Georgieva and Ahring, 2004) and the latter with several subspecies of *Caldanaerobacter subterraneus* (Yokoyama *et al.*, 2009).

Other extremophiles

As was stated earlier, the thermophiles are amongst the best investigated of the extremophiles, but there are many other species which survive under equally challenging environmental conditions and which may also have some potential as the starting point for future methods of reduced pollution manufacturing. For example, cold environments are more common on earth than hot ones. The average oceanic temperature is around 1–3 °C and vast areas of the global land mass are permanently or near-permanently frozen. In these seemingly inhospitable conditions, extremophiles, known as psychrophiles, flourish. A variety of organisms including a number of bacteria and photosynthetic eukaryotes can tolerate these circumstances, often with an optimum functional temperature as low as 4 °C and stopping reproduction above 12 or 15 °C.

Intensely saline environments, such as exist in natural salt lakes or within the artificial confines of constructed salt evaporation ponds are home to a group of extremophiles, termed the halophiles. Under normal circumstances, water flows from areas of low solute concentration to areas where it is higher. Accordingly, in salty conditions, unprotected cells rapidly lose water from their cytoplasm and dehydrate. Halophilic microbes appear to deal with this problem by ensuring that their cytoplasm contains a higher solute concentration than is present in their surroundings. They seem to achieve this by two distinct mechanisms, either manufacturing large quantities of solutes for themselves or concentrating a solute extracted from external sources. A number of species, for example accumulate potassium chloride (KCl) in their cytoplasm, with the concomitant result that extremozymes isolated from these organisms will only function properly in the presence of high KCl levels. By the same token, many surface structural proteins in halophiles require severely elevated concentrations of sodium salts.

Acidophiles thrive in the conditions of low pH, typically below 5, which occur naturally as a result of sulphurous gas production in hydrothermal vents and may also exist in residual spoils from coal mining activity. Though they can tolerate an externally low pH, an acidic intra-cellular environment is intolerable to acidophilic organisms, which rely on protective molecules in, or on, their cell walls, membranes or outer cell coatings to exclude acids. Extremozymes capable of functioning below pH 1 have been isolated from these structures in some acidophile species.

At the other end of the spectrum, alkaliphiles are naturally occurring species which flourish in soda lakes and heavily alkaline soils, typically enduring pH 9 or more. Like the previous acidophiles, alkaliphiles require more typically neutral internal conditions, again relying on protective chemicals on or near their surfaces or in their secretions to ensure the external environment is held at bay.

Diverse degradative abilities

Bacteria possessing pathways involved in the degradation of a number of organic molecules of industrial importance have been acknowledged for some time. One oft quoted example is that for toluene degradation in *Pseudomonas putida*, which exhibits a fascinating interplay between the genes carried on the chromosome and the plasmids (Burlage, Hooper and Sayler, 1989). When a new class of biopolymer produced by the bacterium, *Ralstonia eutropha*, containing sulphur in its backbone, was identified Lűtke-Eversloh *et al.* (2001), it opened up the renewed possibility of other novel biopolymers awaiting discovery, that might have innovative and exciting applications in clean technology. Thus, bacteria are constantly being discovered which exhibit pathways involved in the degradation and synthesis of chemicals of particular interest to environmental biotechnologists.

Within the last ten years, a spate of bacteria representing very diverse degradative abilities have been discovered in a variety of niches adding almost daily, to the pool of organisms of potential use to environmental biotechnology. By illustration these include, phenol degrading *Oceanomonas baumannii* isolated from estuarine mud from the mouth of the River Wear, UK (Brown, Sutcliffe and Cummings, 2001), chloromethane utilising *Hyphomicrobium* and *Methylobacterium* from polluted soil near a petrochemical factory in Russia (McDonald *et al.*, 2001) and a strain of *Clostridium* able to degrade cellulose, isolated from soil under wood chips or the forest floor in North East USA. In addition to their cellulolytic activity, these *Clostridia* were also found to be mesophilic, nitrogen fixing, spore forming and obligate anaerobes. (Monserrate, Leschine and Canale-Parola, 2001) Again, there is interest in this organism with regard to clean technology in the hope that it may be used to convert cellulose into industrially useful substances. A note of caution is that cellulose is a major product of photosynthesis and, being the most abundant biopolymer on this planet, has a vital role to play in the carbon cycle. Large scale disturbance of this balance may have consequences to the environment even less welcome than the technologies they seek to replace. However, judicious use of this biotechnology could reap rewards at many levels.

Bacteria have also adapted to degrade man-made organics called xenobiotics.

Xenobiotics and Other Problematic Chemicals

The word is derived from the Greek '*xenos*' meaning foreign. Throughout this book the definition used is that xenobiotics are compounds which are not produced by a biological procedure and for which no equivalent exists in nature. They present a particular hazard if they are subject to bioaccumulation especially so if they are fat soluble since that enables them to be stored in the body fat of organisms providing an obvious route into the food chain. Despite the fact that these chemicals are man made, they may still be degraded by micro-organisms if they fit into one of the following regimes; gratuitous degradation, a process whereby the xenobiot resembles a natural compound sufficiently closely that it is recognised by the organism's enzymes and may be used as a food source, or cometabolism where the xenobiot is degraded again by virtue of being recognised by the organism's enzymes but in this case its catabolism does not provide energy and so can not be the sole carbon source. Consequently, cometabolism may be sustained only if a carbon source is supplied to the organism. The ability of a single compound to be degraded can be affected by the presence of other contaminants. For example, heavy metals can affect the ability of organisms to grow, the most susceptible being Gram positive bacteria, then Gram negative. Fungi are the most resistant and actinomycetes are somewhere in the middle. This being the case, model studies predicting the rate of contaminant degradation may be skewed in the field where the composition of the contamination may invalidate the study in that application. Soil microorganisms, in particular are very versatile and may quickly adapt to a new food source by virtue of the transmission of catabolic plasmids. Of all soil bacteria, Pseudomonads seem to have the most highly developed ability to adapt quickly to new carbon sources. In bacteria, the genes coding for degradative enzymes are often arranged in clusters, or operons, which usually are carried on a plasmid. This leads to very fast transfer from one bacterium to another especially in the case of *Pseudomonas* where many of the plasmids are self-transmissible. The speed of adaptation is due in part to the exchange of plasmids but in the case of the archaeans particularly, the pathways they carry, which may have been latent over thousands of bacterial generations, owe their existence to previous exposure over millions of years to an accumulated vast range of organic molecules. It is suggested that, unless there has been evolutionary pressure to the contrary, these latent pathways are retained to a large extent requiring little modification if any to utilise new xenobiotics. Even so, bioremediation may require that organisms are altered in some way to make them more suitable for the task and this topic is addressed in Chapter 9. Briefly, the pathways may be expanded by adaptation to the new molecule, or very much less commonly, wholescale insertion of 'foreign' genes may occur by genetic manipulation. There have been several cases reported where catabolic pathways have been expanded in the laboratory. Hedlund and Staley (2001) isolated a strain of *Vibrio cyclotrophicus* from marine sediments contaminated with creosote. By supplying the bacteria with only phenanthrene as a carbon and energy source, the bacteria were trained to degrade several poly aromatic hydrocarbons (PAHs) although some of these only by cometabolism with a supplied carbon source.

Endocrine disrupters

To date, there are chemicals, including xenobiotics, which still resist degradation in the environment. This may be due to a dearth at the site of contamination, of organisms able to degrade them fully or worse, microbial activity which changes them in such a way that they pose a bigger problem than they did previously. One such example is taken from synthetic oestrogens such as 17α-ethinyloestradiol commonly forming the active ingredient of the birth control pills, and the natural oestrogens which, of course, are not xenobiotics. Natural oestrogens are deactivated in humans by glucuronidation, as shown in Figure 3.1, which is a conjugation of the hormone with UDP-glucuronate making the compound more polar and easily cleared from the blood by the kidneys. It is in this modified and inactive form that it is excreted into sewage. However, bacteria present in the aerobic secondary treatment in sewage treatment plants, have the enzyme, *β-glucuronidase*, which removes this modification thus reactivating the hormone.

As an aside, glucuronidation is not confined to hormones but is a process used to detoxify a number of drugs, toxins and carcinogens in the liver. The enzyme catalysing this process is induced in response to prolonged exposure to the toxin thus imparting increased tolerance or even resistance to the chemical.

Returning to the problem of elevated levels of active hormones in the waterways, another aspect is that steroids do not occur in bacteria, although they are present in fungi, and so bacteria lack the necessary pathways to allow complete degradation of these hormones at a rate compatible with the dwell time in sewage treatment plants. The consequence has been raised levels of reactivated oestrogen and 17α-ethinyloestradiol in the waterways leading to disturbances of the endocrine, or hormonal, system in fauna downstream from sewage treatment plants. Such disturbances have been monitored by measuring the presence of the protein vitellogenin (Sole *et al*., 2001) which is a precursor to egg yolk protein, the results of which have indicated feminisation of male fish in many species

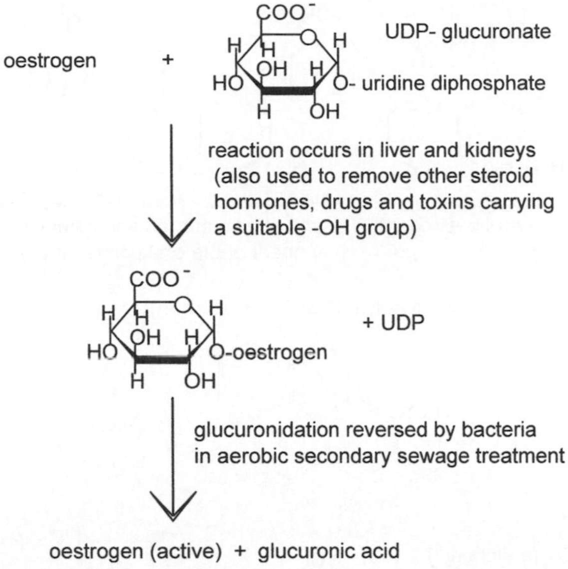

Figure 3.1 *Glucuronidation*

including minnows, trout and flounders. The source of environmental oestrogens is not confined to outfall from sewage treatment plants, however, the fate of endocrine disrupters, examples of which are given in Figure 3.2, in sewage treatment plants is the subject of much research (Byrns, 2001). Many other chemicals, including PAHs, dichlorodiphenyltrichloroethane (DDT), alkyl phenols and some detergents may also mimic the activity of oestrogen. There is general concern as to the ability of some organisms to accumulate these endocrine disrupters in addition to the alarm being raised as to the accumulative effects on humans of oestrogen like activity from a number of xenobiotic sources.

To date there is no absolute evidence of risk to human health but the Environmental Agency and Water UK are recommending the monitoring of environmental oestrogens in sewage treatment outfall. Assays are being developed further to make these assessments (Gutendorf and Westendorf, 2001) and to predict potential endocrine disrupter activity of suspected compounds (Takeyoshi *et al.*, 2002). Oestrogen and progesterone are both heat labile. In addition, oestrogen appears to be susceptible to treatment with ultra violet light, the effects of which are augmented by titanium dioxide (Coleman *et al.*, 2004). The oestrogen is degraded completely to carbon dioxide and water thus presenting a plausible method for water polishing prior to consumption.

Another method for the removal of oestrogens from water, in this case involving *Aspergillus*, has also been proposed (Ridgeway and Wiseman, 1998). Sulphation of the molecule by isolated mammalian enzymes, as a means of hormone inactivation is also being investigated (Suiko *et al.*, 2000). Taken overall, it seems unlikely that elevated levels of oestrogen in the waterways

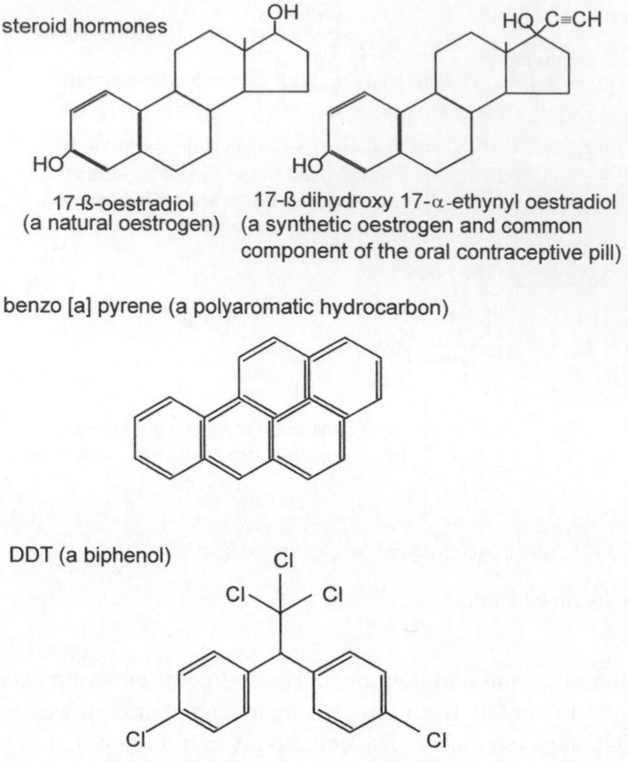

steroid hormones

OH

HO C≡CH

HO

HO

17-ß-oestradiol
(a natural oestrogen)

17-ß dihydroxy 17-α-ethynyl oestradiol
(a synthetic oestrogen and common
component of the oral contraceptive pill)

benzo [a] pyrene (a polyaromatic hydrocarbon)

DDT (a biphenol)

Cl

Cl   Cl

Cl

Cl

Cl

Figure 3.2   *Endocrine disrupters*

will pose a problem to human health in drinking water, although this does not address the problem affecting hormone susceptible organisms living in contaminated water and thus exposed to this potential hazard.

Ongoing discoveries

Almost daily, there are novel bacteria being reported in the literature which have been shown to have the capacity to degrade certain xenobiots. Presumably the mutation which occurred during the evolution of the organism conferred an advantage, and selective pressure maintained that mutation in the DNA, thus producing a novel strain with an altered phenotype. Some examples of such isolates are described here. Reference has already been made to some PAHs mimicking oestrogen which earns those chemicals the title of 'endocrine disrupters'. This is in addition to some being toxic for other reasons and some being carcinogenic or teratogenic. The PAHs are derived primarily from the petrochemicals industry and are polycyclic hydrocarbons of three or more rings which include as members, naphthalene and phenanthrene and historically have been associated with offshore drilling, along with alkylphenols. Several genera of bacteria are now known to be able to degrade PAHs and a novel strain of *Vibrio cyclotrophicus* able to digest naphthalene and phenanthrene, has been isolated from creosote contaminated marine sediments from Eagle Harbour, Washington, USA. It would appear that bacteria isolated from the same marine or estuarine environments may vary quite considerably in their abilities to degrade certain PAHs. This observation is viewed as indicative of diverse catabolic pathways demonstrated by these organisms and awaiting our full understanding (Hedlund and Staley, 2001).

Polycyclic hydrocarbons (PCBs) are xenobiotics which, due to their high level of halogenation, are substrates for very few pathways normally occurring in nature, although a strain of *Pseudomonas putida*, that is able to degrade PCBs has been isolated from waste water outflow from a refinery. This was achieved by the bacterium employing two pathways encoded by two separate operons; the *tod* pathway employed in toluene degradation, and the *cmt* pathway which normally is responsible for the catabolism of *p*-cumate which is a substituted toluene. The mutation which allowed this strain to utilise the *cmt* pathway was found to be a single base change to the promoter-operator sequence. This allowed all the enzymes in this pathway to be expressed under conditions where their synthesis would normally be repressed. Thus, the two pathways could work in conjunction with each other to metabolise PCBs, a relationship described as mosaic (Ohta *et al.*, 2001).

The pthalates are substituted single ring phenols and include terephthalic acid and its isomers, the major chemicals used in the manufacture of polyester fibres, films, adhesives, coatings and plastic bottles. In Chapter 2, and earlier in this chapter, homage has been paid to the resources of genetic capability exhibited by the archaeans. Analysis of anaerobic sewage sludge has revealed a methanogenic consortium of over 100 bacterial clones with the capability to digest terephthalate. Characterisation of these by analysis of their ribosomal DNA sequences, revealed that almost 70% were archaeans, most of which had not been previously identified, and that nearly 90% of the total bacteria comprised two of the novel archaean species. These two species believed to be responsible for the degradation of terephthalic acid (Wu *et al.*, 2001). During wastewater treatment, terepthalic acid is usually treated by aerobic processes, however, this consortium, or others like it provide an anaerobic alternative which, being methanogenic, may be structured to offset processing costs by the utilisation of the methane.

Mobility of DNA

Throughout this book, reference is made to the movement of genes within and between organisms. The reason why it appears at all in a book on environmental biotechnology is to emphasise the 'oneness' of the environment, not just at the more obvious level of industrial impact but right down to the interaction between the genetic material of organisms themselves. Plasmids may be transferred between bacteria by conjugation of which there are several types, but all of which require direct cell to cell contact. Not only are genes transferred between bacteria on plasmids, but bacteriophages (bacterial viruses) are also vectors for intercellular transmission. Similarly, eukaryotic viruses are able to transfer genetic material between susceptible cells. In addition, bacterial cells may pick up DNA free in the environment under conditions where their cell wall has become 'leaky' to fragments of this macromolecule, a process called transformation. There is also considerable rearrangement of genomic material within an organism stimulated by the presence of transposons. There are many classes of transposable elements which are short pieces of DNA, able to excise themselves, or be excised, out of a genome. Often they take with them neighbouring pieces of DNA, and then reinsert themselves, sometimes with the assistance of other genes, into a second site distinct from the original location on the same genome. Insertion may be into specific sites or random depending on the nature of the transposon. Transposition normally requires replication of the original DNA fragment and so a copy of this transposon is transferred leaving the original behind. Transposition is widespread and occurs in virtually all organisms for which evidence of this process has been sought, both prokaryotic and eukaryotic. The term 'transposable element', was first coined by Barbara MacClintock, who discovered them in maize, publishing her data in the early 1950s. However, it was not until many years later that the full significance of her work was being recognised, with similar elements being discovered in bacteria. Transposable elements are known to promote the fusion of plasmids within a bacterial cell, where more than one type of plasmid is present. The fusion is stimulated by the presence of insertion sequences (ISs), which are short pieces of DNA of a defined and limited range of sequences. They are often found at either end of a transposable element. Their presence enables various DNA rearrangements to take place leading to moderation of gene expression. Taking together the reorganisation of DNA within all types of organisms attributable to transposable elements and IS, with transfer of DNA between organisms by plasmids and transformation, in the case of prokaryotes, and viruses in the case of both prokaryotes and eukaryotes, the potential for DNA rearrangement within and between organisms is enormous.

It has been proposed (Reanney, 1976), that such transfer is far more universal than had previously been voiced. Transfer of genes by extra chromosomal elements (ECEs), which is the all embracing name given to include plasmids and viruses, models the means by which molecular evolution takes place in the environment. The proposal is that the evolutionary process occurs principally by insertions and deletions of the genome such as those caused by the activities of ECEs and transposable elements and not by point mutations more frequently observed in isolated cultures such as those maintained in laboratory conditions. It is further suggested that much of the phenotypic novelty seen in evolution is the result of rearrangement of existing structural genes into a different region of the genome and therefore operating under different parameters affecting gene regulation. Transfer of genes across wide taxonomic gaps is made possible by the mobile nature of ECEs, many of which may cross species barriers often resulting in the insertion of all or part of the ECE into the recipient genome.

Examples of such mobility are viruses which infect a wide host range, such as some retroviruses, the alfalfa mosaic virus and the Ti plasmid of *Agrobacterium tumefaciens* which the bacterium introduces into plant cells. The retroviruses, of which Human Immunodeficiency Virus (HIV) is an example, are unusual in having RNA as their genetic material. They replicate in a manner which includes double stranded DNA as an intermediate and so may integrate into the host cell genome. RNA viruses tend to be more susceptible than DNA viruses to mutation presumably due to the less chemically stable nature of the macromolecule. They have been invoked as being the likely agents for the spread of genetic information between unrelated eukaryotes by Reanney (1976). His observations led him to conclude that there is only a blurred distinction between cellular and ECE DNA both in eukaryotes and prokaryotes and further suggests that no organism lives in true genetic isolation as long as it is susceptible to at least one of the classes of ECEs described above. Clearly, for the mutation to be stabilised, it must occur in inheritable DNA sequences, a situation reasonably easy to achieve in microbes and at least possible in multicellular organisms.

The existence of genetic mobility has been accepted for many years, even though the extent and the mechanisms by which it operates are still being elucidated. From this knowledge several lessons may be learned; among them, that the genetic environment of any organism may well be significant and that there is some justification in viewing the principle of genetic engineering as performing in the laboratory, a process which is occurring in abundance throughout the living world. This is a topic, that is further explored in later chapters.

Closing Remarks

As has been seen, even within the brief discussion in this chapter, life on Earth is a richly varied resource and the functional reality of biodiversity is that many more metabolic pathways exist, particularly within the microbial melting pot, than might be commonly supposed. As a result, a number of generally unfamiliar groups of chemicals and organisms have implications for the application of environmental biotechnology which exceed their most obvious contributions to a wider consideration of the life sciences. Hence, xenobiotics, as an example of the former, represent a current problem for which the solution remains largely unresolved and extremophiles, as the latter, hold the potential to revolutionise many industrial procedures, thereby heralding major benefits in terms of 'clean technology'. There are many aspects of current environmental management for which there is no presently relevant biotechnological intervention. However, this is not a static science, either in terms of what can be done, or the tools available. Thus, while the bulk of the rest of this book addresses the sorts of biotechnological applications that have now become fairly routinely applied to environmental problems, discoveries and developments both within the field and from other disciplines can and do filter in and alter the state of the possible.

References

Brown, G.R., Sutcliffe, I.C. and Cummings, S.P. (2001) Reclassification of [Pseudomonas] *doudoroffi* (Baumann *et al.* 1983) into the genus *Oceanomonas* gen. nov. as *Oceanomonas doudoroffi* comb. nov., and description of a phenol-degrading bacterium from estuarine water as *Oceanomonas baumanni* sp. nov. *International Journal of Systemic and Evolutionary Microbiology*, **51**, 67–72.

Burlage, R.S., Hooper, S.W. and Sayler, G.S. (1989) The Tol (PWWO) catabolic plasmid. *Applied and Environmental Microbiology*, **55**, 1323–1328.

Byrns, G. (2001) The fate of xenobiotic organic compounds in wastewater treatment plants. *Water Research*, **35**, 2523–2533.

Coleman, H.M., Routledge, E. J., Sumpter, J. P., Eggins, B. R. and Byrne, J. A. (2004) Rapid loss of estrogenicity of steroid estrogens by UVA photolysis and photocatalysis over an immobilised titanium dioxide catalyst. *Water Research*, **38**(14–15), 3233–3240.

Gutendorf, B. and Westendorf, J. (2001) Comparison of an array of in vitro assays for the assessment of the estrogenic potential of natural and synthetic estrogens, phytoestrogens and xenoestrogens. *Toxicology*, **166**, 79–89.

Hedlund, B.P. and Staley, J.T. (2001) *Vibrio cyclotrophicus* sp. nov., a polycyclic aromatic hydrocarbon (PAH) – degrading marine bacterium. *International Journal of Systemic and Evolutionary Microbiology*, **51**, 61–66.

Lütke-Eversloh, T., Bergander, K., Luftmann, H. and Steinbüchel, H. (2001) Identification of a new class of biopolymer: bacterial synthesis of a sulphur-containing polymer with thioester linkages. *Microbiology*, **147**, 11–19.

McDonald, I.R., Doronina, N.V., Trotsenko, Y.A. *et al.* (2001) *Hyphomicrobium chloromethanicum* sp. nov. and *Methylobacterium chloromethanicum* sp. nov., chloromethane-utilizing bacteria isolated from a polluted environment. *International Journal of Systemic and Evolutionary Microbiology*, **51**, 119–122.

Monserrate, E., Leschine, S.B. and Canale-Parola, E. (2001) *Clostridium hungatei* sp. nov., a mesophilic, N_2-fixing cellulolytic bacterium isolated from soil. *International Journal of Systematic and Evolutionary Microbiology*, **51**, 123–132.

Ohta, Y., Maeda, M. and Kudo, T. (2001) *Microbiology*, **147**, 31–41.

Reanney, D. (1976) Extrachromosomal elements as possible agents of adaptation and development. *Bacteriological Reviews*, **40**, 552–90.

Ridgeway, T.J. and Wiseman, H. (1998) *Biochemical Society Transactions*, **26** (4): 675–80.

Sole, M., Porte, C. and Barcelo, D. (2001) Analysis of the estrogenic activity of sewage treatment works and receiving waters using vitellogenin induction in fish as a biomarker. *TRAC-Trends in Analytical Chemistry*, **20**, 518–525.

Sommer, P., Georgieva, T. and Ahring, B.K. (2004) Potential for using thermophilic anaerobic bacteria for bioethanol production from hemicelluloses. *Biochemical Society Transactions*, **32** (Pt 2), 283–289.

Suiko, M, Sakakibara, Y, and Liu, M.C. (2000) Sulfation of environmental of estrogen-like chemicals by human cytosolic sulfotransferases. *Biochemical and Biophysical Research Communications*, **267** (1): 80–4.

Takeyoshi, M., Yamasaki, K., Sawaki, M. *et al.* (2002) The efficacy of endocrine disrupter screening tests in detecting anti-estrogenic receptor-ligand effects downstream of interactions. *Toxicology Letters*, **126**, 91–98.

Wright, P.C. and Raper, J.A. (1996) A review of some parameters involving fluidized bed bioreactors. *Chemical Engineering Technology*, **19**, 50–64.

Wu, J.H., Liu, W.T., Tseng, I.C. and Cheng, S.S. (2001) Characterization of microbial consortia in a terephthalate-degrading anaerobic granular sludgesystem. *Microbiology*, **147**, 373–82.

Yokoyama, H., Ohmori, H., Waki, M. *et al.* (2009) Continuous hydrogen production from glucose by using extreme thermophilic anaerobic microflora. *Journal of Bioscience and Bioengineering*, **107** (1), 64–66.

Case Study 3.1 Endocrine Disruptors in the Aquatic Environment (Switzerland)

Although the potential for endocrine disruptors in the environment to adversely affect humans, other animals and even entire ecosystems is well recognised, the chemical and micro-pollutant flows within specific water courses has generally been less well addressed. Recognising the need for such systematic analysis and modelling to enable meaningful quality standards to be drawn up and appropriate impact-related action plans to be implemented, the Swiss National Science Foundation undertook a major research programme to investigate the issue, under the auspices of the country's Federal Council.

The study found that endocrine disruptors presented a local and regional problem for the aquatic environment within Switzerland. This was particularly apparent in waters receiving domestic wastewater flows from treatment facilities either located in densely populated regions or where discharge entered small streams and the dilution effect was limited (<1 : 10).

A significant part of the work focused on the oestrogenic feminisation of fish, examining the incidence of increased vitellogenin (Vg) levels in males as a key bio-indicator. Vg is a precursor protein for egg yolk formation and thus normally dormant in males, but when they are exposed to natural or synthetic oestrogens, the Vg gene is expressed in a dose-dependent manner, making it a useful molecular marker for exposure. The study revealed that elevated Vg was only a feature of a minority of the water bodies investigated and then only in 'slight to moderate' amounts. Perhaps most interestingly, there were no indications that the high levels of gonadal irregularities reported in Lake Thun whitefish could be attributed to endocrine disruptors, which obviously highlights the importance of such comprehensive studies in both long-term catchment area management and effective environmental regulation.

Laboratory experiments have established that chronic exposure to endocrine disruptors at environmentally relevant concentrations can produce developmental, growth and reproductive problems in a range of aquatic creatures, with fish

Continued on page 64

Continued from page 63

displaying effects even when exposed to very low levels. As a result, endocrine disruptors – and synthetic oestrogens, in particular – represent one of the clearest examples of the significant challenge xenobiotic contamination poses for environmental management.

4

Pollution and Pollution Control

Pollution has become one of the most frequently talked about of all environmental problems by the world at large and yet, in many respects, it can often remain one of the least understood. The word itself has a familiar ring to it and inevitably the concept of pollution has entered the wider consciousness as a significant part of the burgeoning 'greening' of society in general. However, the diverse nature of potentially polluting substances can lead to some confusion. It is important to realise that not all pollutants are manufactured or synthetic, that under certain circumstances, many substances may contribute to pollution and that, perhaps most importantly for our purposes, any biologically active substance has the potential to give rise to a pollution effect. This inevitably leads to some difficulty in any attempt at classifying pollutants, since clearly, they do not represent a single unified class, but rather a broad spectrum. While it is possible, as we shall discuss shortly, to produce a means of systematic characterisation of pollutant substances, though useful for a consideration of wider contamination effects, this is an inherently artificial exercise. It is, therefore, perhaps more useful to begin the discussion with a working definition and although it is now more than two decades old, the statutory version contained in the UK Environmental Protection Act, 1990 (EPA) remains as good as any.

> 'Pollution of the environment' means pollution of the environment due to the release (into any environmental medium) from any process of substances which are capable of causing harm to man or any other living organisms supported by the environment.
>
> *EPA, Introduction.*

> ....the escape of any substance capable of causing harm to man or any other living organism supported by the environment.
>
> *EPA, Section 29, Part II.*

In essence, then, pollution is the introduction of substances into the environment which, by virtue of their characteristics, persistence or the quantities involved, are likely to be damaging to the health of humans, other animals and plants, or otherwise compromise that environment's ability to sustain life. It should be obvious that this is an expressly inclusive definition, encompassing

Environmental Biotechnology: Theory and Application, Second Edition Gareth M. Evans and Judith C. Furlong
© 2011 John Wiley & Sons, Ltd.

not simply the obviously toxic or noxious substances, but also other materials which can have a polluting effect under certain circumstances.

Classifying Pollution

While, as we said earlier, this diverse nature of potential pollutants makes their systematisation difficult in absolute terms, it is possible to produce functional classifications on the basis of various characteristics. However, it must be clearly born in mind that all such classification is essentially artificial and subjective, and that the system to be adopted will typically depend on the purpose for which it is ultimately intended. Despite these limitations, there is considerable value in having some method, if only as a predictive environmental management tool, for considerations of likely pollutant effect.

Classification may, for example be made on the basis of the chemical or physical nature of the substance, its source, the environmental pathway used, the target organism affected or simply its gross effect. Figure 4.1 shows one possible example of such a categorisation system and clearly many others are possible.

The consideration of a pollutant's properties is a particularly valuable approach when examining real-life pollution effects, since such an assessment requires both the evaluation of its general properties and the local environment. This may include factors such as:

- toxicity;
- persistence;
- mobility;
- ease of control;
- bioaccumulation;
- chemistry.

Toxicity

Toxicity represents the potential damage to life and can be both short and long term. It is related to the concentration of pollutant and the time of exposure to it, though this relationship is not an easy one. Intrinsically highly toxic substances

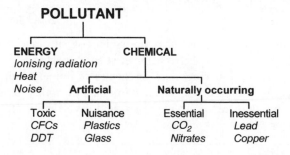

Figure 4.1   *Pollution classification*

can kill in a short time, while less toxic ones require a longer period of exposure to do damage. This much is fairly straightforward. However, some pollutants which may kill swiftly in high concentrations, may also have an effect on an organism's behaviour or its susceptibility to environmental stress over its lifetime, in the case of low concentration exposure.

Availability also features as an important influence, both in a gross, physical sense and also in terms of its biological availability to the individual organism, together with issues of its age and general state of health. Other considerations also play a significant part in the overall picture of toxicity and we shall return to look at some of them in greater depth shortly.

Persistence

This is the duration of effect. Environmental persistence is a particularly important factor in pollution and is often linked to mobility and bioaccumulation.

Highly toxic chemicals which are environmentally unstable and breakdown rapidly are less harmful than persistent substances, even though these may be intrinsically less toxic.

Mobility

The tendency of a pollutant to disperse or dilute is a very important factor in its overall effect, since this affects concentration. Some pollutants are not readily mobile and tend to remain in 'hot-spots' near to their point of origin. Others spread readily and can cause widespread contamination, though often the distribution is not uniform. Whether the pollution is continuous or a single event, and if it has arisen from a single point or multiple sources, form important considerations.

Ease of control

Many factors contribute to the overall ease with which any given example of pollution can be controlled, including the mobility of the pollutant, the nature, extent or duration of the pollution event and local site-specific considerations. Clearly, control at source is the most effective method, since it removes the problem at its origin. However, this is not always possible and in such cases, containment may be the solution, though this can itself lead to the formation of highly concentrated hot-spots.

For some substances, the dilute and disperse approach, which is discussed more fully later in this chapter, may be more appropriate, though the persistence of the polluting substances must obviously be taken into account when making this decision.

Bioaccumulation

As is widely appreciated, some pollutants, even when present in very small amounts within the environment, can be taken up by living organisms and become concentrated in their tissues over time. This tendency of some chemicals to be taken up and then concentrated by living organisms is a major consideration, since even relatively low background levels of contamination may accumulate up the food chain.

Chemistry

Pollution effects are not always entirely defined by the initial nature of the contamination, since the reaction or breakdown products of a given pollutant can sometimes be more dangerous than the original substance. This is of particular relevance to the present discussion, since the principle underlying much of practical bioremediation in general involves the break down of pollutants to form less harmful products.

This is further complicated in that while the chemistry of the pollutant itself is clearly important, other substances present and the geology of the site may also influence the outcome. Accordingly, both synergism and antagonism are possible. In the former, two or more substances occurring together produce a combined pollution outcome which is greater than simply the sum of their individual effects; in the latter, the combined pollution outcome is smaller than the sum of each acting alone.

The Pollution Environment

There is sometimes a tendency for contamination to be considered somewhat simplistically, in isolation from its context. It is important to remember that pollution cannot properly be assessed without a linked examination of the environment in which it occurs. The nature of the soil or water which harbours the pollution can have a major effect on the actual expressed end-result. In the case of soil particularly, many properties may form factors in the modification of the contamination effect. Hence, the depth of soil, its texture, type, porosity, humus content, moisture, microbial complement and biological activity can all have a bearing on the eventual pollution outcome. Moreover, recent work on the translatory flow of water in soil has shown that under certain circumstances, particularly after prolonged dry spells, the soil matrix can hold novel precipitation so tightly that it almost never mixes with other water (Brooks *et al.*, 2010). Inevitably, the combination of all these factors can make accurate prediction difficult, though a consideration of system stability can often give a good indication of the most likely pollution state of a given environment.

The more stable and robust the environmental system affected, the less damage a given pollution event will inflict and clearly, fragile ecosystems or sensitive habitats are most at risk. It should be obvious that, in general terms, the post-pollution survival of a given environment depends on the maintenance of its natural cycles. Equally obviously, artificial substances which mimic biological molecules can often be major pollutants since they can modify or interrupt these processes and pollution conversion can spread or alter the effect.

Pollution Control Strategies

Dilution and dispersal

The concept of 'dilute and disperse' was briefly mentioned earlier in this discussion. In principle, it involves the attenuation of pollutants by permitting them to become physically spread out, thereby reducing their effective point concentration. The dispersal and the consequent dilution of a given substance depends on its nature and the characteristics of the specific pathway used to achieve this. It may take place, with varying degrees of effectiveness, in air, water or soil.

Air

In general terms, air movement gives good dispersal and dilution of gaseous emissions. However, heavier particulates tend to fall out near the source and the mapping of pollution effects on the basis of substance weight/distance travelled is widely appreciated.

Water

Typically, there is good dispersal and dilution potential in large bodies of water or rivers, but smaller watercourses clearly have a correspondingly lower capacity. It is also obvious that moving bodies of water disperse pollutants more rapidly than still ones.

Soil

Movement through the soil represents another opportunity for the dilute and disperse approach, often with soil water playing a significant part, except where excessive ecohydrologic separation becomes a limiting factor (Brooks *et al.*, 2010), and typically aided by the activities of resident flora and fauna. The latter generally exerts an influence in this context which is independent of any bioaccumulation potential.

Concentration and containment

The principle behind this is diametrically opposed to the previous approach, in that instead of relying on the pollutant becoming attenuated and spread over a wide area, it is an attempt to gather together the offending substance and prevent its escape into the surrounding environment.

The inherent contradiction between these two general methods is an enduring feature of environmental biotechnology and, though the fashion changes from time to time, favouring first one and then the other, it is fair to say that there is a place for both, dependent on individual circumstances. As with so much relating to the practical applications of biotechnologies to environmental problems, the idea of a 'best' method, at least in absolute terms, is of little value. The whole issue is far more contextually sensitive and hence the specific modalities of the particular, are often more important concerns than the more theoretically applicable general considerations.

Practical Toxicity Issues

The general factors which influence toxicity have already been set out earlier in this discussion, but before moving on to consider wider practical issues it is helpful to mention briefly the manner in which the toxic action of pollutants arises. There are two main mechanisms, often labelled 'direct' and 'indirect'. In the former, the effect arises by the contaminant combining with cellular constituents or enzymes and thus preventing their proper function. In the latter, the damage is done by secondary action resulting from their presence, typified by histamine reactions in allergic responses.

The significance of natural cycles to the practical applications of environmental biotechnology is a point that has already been made. In many respects the functional toxicity of a pollution event is often no more than the obverse aspect of this same coin, in that it is frequently an overburdening of existing innate systems which constitutes the problem. Thus the difficulty lies in an inability to deal with the contaminant by normal routes, rather than the simple presence of the substance itself. The case of metals is a good example. Under normal circumstances, processes of weathering, erosion and volcanic activity lead to their continuous release into the environment and corresponding natural mechanisms exist to remove them from circulation, at a broadly equivalent rate. However, human activities, particularly after the advent of industrialisation, have seriously disrupted these cycles in respect of certain metals, perhaps most notably cadmium, lead, mercury and silver. While the human contribution is, clearly, considerable, it is also important to be aware that there are additional potential avenues of pollution and that other metals, even though natural fluxes remain their dominant global source, may also give rise to severe localised contamination at times.

The toxicity of metals is related to their place in the periodic table, as shown in Table 4.1 and reflects their affinity for amino and sulphydral groups (associated with active sites on enzymes).

In broad terms, type-A metals are less toxic than type-B, but this is only a generalisation and a number of other factors exert an influence in real-life situations.

Passive uptake by plants is a two-stage process, beginning with an initial binding onto the cell wall followed by diffusion into the cell itself, along a

Table 4.1   *Metal periodicity and toxicity*

Metal group	Relative toxicity
Group IA	Na < K < Rb and Cs
Group IB	Cu < Ag < Au
Group IIA	Mg < Ca < Sr < Ba
Group IIB	Zn < Cd < Hg
Group IIIA	Al < Ga < In < Tl

concentration gradient. As a result, those cations which readily associate with particulates are accumulated more easily than those which do not. In addition, the presence of chelating ligands may affect the bio-availability and thus, the resultant toxicity of metals. Whereas some metal–organic complexes (Cu–EDTA, for example) can detoxify certain metals, lipophilic organometallic complexes can increase uptake and thereby the functional toxic effect observed.

Although we have been considering the issue of metal toxicity in relation to the contamination of land or water, it also has relevance elsewhere and may be of particular importance in other applications of biotechnologies to environmental problems. For example, anaerobic digestion is an engineered microbial process commonly employed in the water industry for sewage treatment and gaining acceptance as a method of biowaste management. The effects of metal cations within anaerobic bioreactors are summarised in Table 4.2, and from which it is apparent that concentration is the key factor.

However, the situation is not entirely clear cut as the interactions between cations under anaerobic conditions may lead to increased or decreased effective toxicity in line with the series of synergistic/antagonistic relationships shown in Table 4.3.

Toxicity is often dependent on the form in which the substance occurs and substances forming analogues which closely mimic the properties of essential chemicals are typically readily taken up and/or accumulated. Such chemicals are often particularly toxic as the example of selenium illustrates.

Table 4.2   *The effect of metal cations on anaerobic digestion*

Cation	Stimulatory	Moderately inhibitory	Strongly inhibitory
Sodium	100–200	3 500–5 500	8 000
Potassium	200–400	2 500–4 500	12 000
Calcium	100–200	2 500–4 500	8 000
Magnesium	75–150	1 000–1 500	3 000

Concentrations in mg/l.

Table 4.3   *Effective toxicity and synergistic/antagonistic relationships*

Toxic cation	Synergistic	Antagonistic
Ammonium-N	Calcium Magnesium Potassium	Sodium
Calcium	Ammonium-N Potassium	Sodium Magnesium
Magnesium	Ammonium-N Calcium	Sodium Potassium
Potassium		Sodium
Sodium	Ammonium-N Calcium	Potassium

Often wrongly referred to as a toxic metal, and though it has some metallic properties, selenium is a non-metal of the sulphur group. It is an essential trace element and naturally occurs in soils, though in excess it can be a systemic poison with the LD_{50} for certain selenium compounds being as low as 4 μg/kg body weight.

In plants, sulphur is actively taken up in the form of sulphate $SO_4{}^{2-}$. The similarity of selenium to sulphur leads to the existence of similar forms in nature, namely:

- Selenite $SeO_3{}^{2-}$;
- Selenate $SeO_4{}^{2-}$.

As a result, selenium can be taken up in place of sulphur and become incorporated in normally sulphur containing metabolites.

Practical Applications to Pollution Control

In the next chapter contaminated land and bioremediation, which typically form a wider area of concern for environmental biotechnology will be considered, in some detail.

To give a practical context with which to close this section, however, a brief discussion of air pollution and odour control follows.

Bacteria normally live in an aqueous environment which clearly presents a problem for air remediation. Frequently the resolution is to dissolve the contaminant in water, which is then subjected to bioremediation by bacteria, as in the following descriptions. However, there is scope for future development of a complementary solution utilising the fact that many species of yeast produce aerial hyphae which may be able to metabolise material directly from the air.

A variety of substances can be treated, including volatile organic carbon containing compounds (VOCs) like alcohols, ketones or aldehydes and odorous substances like ammonia and hydrogen sulphide (H_2S). While biotechnology is often thought of as something of a new science, the history of its application to air-borne contamination is relatively long. The removal of H_2S by biological means was first discussed as long ago as 1920 and the first patent for a truly biotech-based method of odour control was applied for in 1934. It was not until the 1960s that the real modern upsurge began, with the use of mineral soil filter media and the first true biofilters were developed in the succeeding decade. This technology, though refined, remains in current use. The latest state-of-the-art developments have seen the advent of the utilisation of mixed microbial cultures to degrade xenobiotics, including chlorinated hydrocarbons like dichloromethane and chlorobenzene.

There are a number of general features which characterise the various approaches applied to air contamination. Typically systems run at an operational temperature within a range of 15–30 °C, in conditions of abundant moisture, at a pH between 6 and 9 and with high oxygen and nutrient availability. In addition, most of the substances which are commonly treated by these systems are water soluble.

The available technologies fall naturally into three main types, namely, biofilters, biotrickling filters and bioscrubbers. To understand these approaches, it is probably most convenient to adopt a view of them as biological systems for the purification of waste or exhaust gases. All three can treat a wide range of flow rates, ranging from 1000 to 100 000 m^3/h, hence the selection of the most appropriate technology for a given situation is based on other criteria. The concentration of the contaminant, its solubility, the ease of process control and the land requirement are, then, principal factors and they interact as shown in Table 4.4 to indicate the likely best approach.

Biofilters

As mentioned earlier, these were the first methods to be developed. The system, shown schematically in Figure 4.2, consists of a relatively large vessel or container, typically made of cast concrete, metal or durable plastic, which holds a filter medium of organic material such as peat, heather, bark chips and the like. The gas to be treated is forced, or drawn, through the filter, as shown in the diagram. The medium offers good water holding capacity and soluble chemicals within the waste gas, or smell dissolve into the film of moisture around the matrix. Bacteria, and other micro-organisms present, degrade components of

Table 4.4   *Odour control technology selection table*

Technology	Compound concentration	Compound solubility	Process control	Land requirement
Biofilter	Low	Low	Low	High
Biotrickling filter	Low-medium	Low-high	Medium-high	Low
Bioscrubber	Low-medium	Medium-high	High	Low-medium

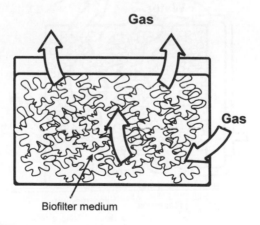

Gas

Gas

Biofilter medium

Figure 4.2   *Biofilter*

the resultant solution, thereby bringing about the desired effect. The medium itself provides physical support for microbial growth, with a large surface area to volume ratio, high in internal void spaces and rich in nutrients to stimulate and sustain bacterial activity. Biofilters need to be watered sufficiently to maintain optimum internal conditions, but waterlogging is to be avoided as this leads to compaction, and hence, reduced efficiency. Properly maintained, biofilters can reduce odour release by 95% or more.

Biotrickling filters

As shown in Figure 4.3, in many respects these represent an intermediate technology between biofilters and bioscrubbers, sharing certain features of each. Once again, an engineered vessel holds a quantity of filter medium, but in this case, it is an inert material, often clinker or slag. Being highly resistant to compaction, this also provides a large number of void spaces between particles and a high surface area relative to the overall volume of the filter. The microbes form an attached growth biofilm on the surfaces of the medium. The odourous air is again forced through the filter, while water simultaneously recirculates through it, trickling down from the top, hence the name. Thus a counter-current flow is established between the rising gas and the falling water, as shown in the diagram, which improves the efficiency of dissolution. The biofilm communities feed on substances in the solution passing over them, biodegrading the constituents of the smell.

Process monitoring can be achieved relatively simply by directly sampling the water recirculating within the filter vessel. Process control is similarly straight-forward, since appropriate additions to the circulating liquid can be made, as

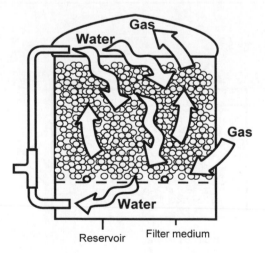

Figure 4.3   *Biotrickling filter*

required, to ensure an optimum internal environment for bacterial action. Though the efficiency of the biotrickling filter is broadly similar to the previous method, it can deal with higher concentrations of contaminant and has a significantly smaller foot-print than a biofilter of the same throughput capacity. However, as with almost all aspects of environmental biotechnology, these advantages are obtained by means of additional engineering, the corollary of which is, inevitably, higher capital and running costs.

Bioscrubbers

Although it is normally included in the same group, the bioscrubber (Figure 4.4) is not itself truly a biological treatment system, but rather a highly efficient method of removing odour components by dissolving them. Unsurprisingly, then, it is most appropriate for hydrophilic compounds like acetone or methanol.

The gas to be treated passes through a fine water spray generated as a mist or curtain within the body of the bioscrubber vessel. The contaminant is absorbed into the water, which subsequently pools to form a reservoir at the bottom. The contaminant solution is then removed to a secondary bioreactor where the actual process of biodegradation takes place. In practice, activated sludge systems (which are described in detail in Chapter 6) are often used in this role.

As in the preceding case, process control can be achieved by monitoring the water phase and adding nutrients, buffers or fresh water as appropriate.

Other options

It is important to be aware that biotechnology is not the only answer to controlling air pollution. A number of alternative approaches exist, though it is clearly beyond the scope of this book to discuss them at length. The following brief outline may

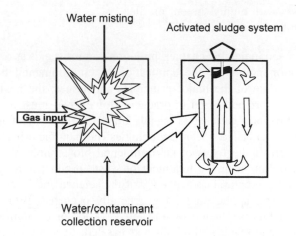

Figure 4.4   *Bioscrubber*

help to give a flavour of the wider context, but to understand how the various technologies compare, the reader should seek more detailed information.

Absorption

Absorbing the compound in a suitable liquid; this may oxidise or neutralise it in the process.

Adsorption

Activated carbon preferentially adsorbs organic molecules; this can be tailored to give contaminant-specific optimum performance.

Incineration

High temperature oxidation; effective against most contaminants, but costly.

Ozonation

Use of ozone to oxidise some contaminants, like hydrogen sulphide; effective but can be costly.

The main advantages of biotechnological approaches to the issue of air contamination can be summarised as:

- competitive capital costs;
- low running costs;
- low maintenance costs;
- low noise;
- no carbon monoxide production;
- avoids high temperature requirement or explosion risk;
- safe processes with highly 'green' profile;
- robust and tolerant of fluctuation.

As was discussed in the first chapter, pollution control stands as one of the three major intervention points for the application of environmental biotechnology. Having defined some of the major principles and issues, the next chapter will examine how they are addressed in practice. However, it must not be forgotten that, as with all tripods, each leg is equally important; the potential contribution to be made by the so-called 'clean technologies' in manufacturing should not be overlooked. Much of the focus of environmental biotechnology centres on the remediation of pollution or the treatment of waste products. In many respects, this tends to form the natural constituency of the science and is, certainly, where the bulk of practical applications have generally occurred. While the benefits of the controlled biodegradation of unwanted wastes or contaminants is clear, this does typify 'end-of-pipe' thinking and has led, to some extent justifiably, to the criticism that it merely represents moving the problem from one place to another. Another option to deal with both these ongoing problems is, simply, to avoid their production in the first place and while this may seem over-idealistic in some aspects, it does have a clear and logical appeal. Throughout this book, 'environmental' biotechnology is defined in the broad sense of the utilisation of applied biological methods to the benefit of the environment. Thus, any use of the life sciences which removes, remediates or obviates contamination of the biosphere falls firmly within its remit and *a priori* action, to avoid the problem in the first place must be preferential. The proverbial ounce of prevention is worth a pound of cure.

The current emphasis on clean-up and treatment is largely the result of his-torical circumstance. As legislation has become more stringent, the regulation of waste and pollution has correspondingly forced the pace of environmental inter-vention. In addition, the prevalence of 'the polluter pays' principle, coupled with ever greater pressures to redevelop existing 'brown-field' sites, in preference to *de novo* development has inevitably necessitated a somewhat reactive response. However, increasingly biotechnologies are being developed which, though per-haps not 'environmental' in themselves, bring significant benefits to this sphere. Their advantages to industry in terms of reduced demands for integrated pollution control and minimised waste disposal costs also suggest a clear likelihood of their success in the commercial sphere. Generally, the environment has tended to fare best when its interests and economic ones go hand in hand and the pre-emptive approach which the new technologies herald seems ideally suited to both.

'Clean' Technology

The mechanisms by which pollution or waste may be reduced at source are varied. They may involve changes in technology or processes, alteration in the raw materials used or a complete restructuring of procedures. Generally speaking, biotechnological interventions are principally limited to the former aspects, though they may also prove instrumental in permitting procedural change. The main areas in which biological means may be relevant fall into three broad categories:

- process changes;
- biological control;
- bio-substitutions.

In the following discussions of these three groups, it is not suggested that the examples cited are either comprehensive or exhaustive; they are simply intended to illustrate the wide potential scope of applications open to biotechnology in clean manufacturing. For precisely the reasons mentioned in respect of the economic aspects of this particular area of industrial activity, the field is a fast evolving one and many more types of biotechnological interventions are likely in the future, especially where commercial pressures derive a competitive advantage.

Process Changes

Replacement of existing chemical methods of production with those based on microbial or enzyme action is an important potential area of primary pollution prevention and is one role in which the use of genetically modified organisms could give rise to significant environmental benefit. Biological synthesis, either by whole organism or by isolated enzymes, tends to operate at a lower temperature and, as a result of high enzymatic specificity, gives a much purer yield with fewer by-products, thus saving the additional cost of further purification. There are many examples of this kind of industrial usage of biotechnology. In the cosmetics sector, there is a high demand for isopropyl myristate which is used in moisturising creams. The conventional method for its manufacture has a large energy requirement, since the process runs at high temperature and pressure to give a product which needs further refinement before it is suitable for use. An alternative approach, using enzyme-based esterification offers a way to reduce the overall environmental impact by deriving a cleaner, odour-free product and at higher yields, with lower energy requirements and less waste for disposal.

Textile industry

There is a long tradition of the use of biological treatment methods in the clothing and textile industry, dating back to the first use of amylase enzymes from malt extract, at the end of the nineteenth century, to degrade starch-based sizes for cheap and effective reduction of fabric stiffness and improvement to its drape. Currently, novel enyzmatic methods provide a fast and inexpensive alternative to traditional flax extraction by breaking down the woody material in flax straw, reducing the process time from 7 to 10 days, down to a matter of hours. The enzyme-based retting processes available for use on hemp and flax produce finer, cleaner fibres, and, consequently, novel processing techniques are being developed to take advantage of this. Interest is growing in the production of new, biodegradable polymeric fibres which can be synthesised using modified soil bacteria, avoiding the current persistence of these materials in landfills, long after garments made from them are worn out.

In natural fibre production enzymes are useful to remove the lubricants which are introduced to prevent snagging and reduce thread breakage during spinning, and to clean the natural sticky secretions present on silk. The process of bio-scouring for wool and cotton uses enzymes to remove dirt rather than traditional chemical treatments and bio-bleaching uses them to fade materials, avoiding both the use of caustic agents and the concomitant effluent treatment problems such conventional methods entail. Biological catalysts have also proved effective in shrink-proofing wool, improving quality while ameliorating the wastewater produced and reducing its treatment costs, compared with chemical means.

A process which has come to be called bio-polishing involves enzymes in shearing off cotton microfibres to improve the material's softness and the drape and resistance to pilling of the eventual garments produced.

Bio-stoning has been widely adopted to produce 'stone-washed' denim, with enzymes being used to fade the fabric rather than the original pumice stone method, which had a higher water consumption and caused abrasion to the denim.

However, perhaps the most fitting example of environmental biotechnology in the textile industry, though not really in a 'clean technology' role, is the incorporation of adsorbers and microbes within a geotextile produced for use in land management around railways. Soaking up and subsequently biodegrading diesel and grease, the textile directly reduces ground pollution, while also providing safer working conditions for track maintenance gangs and reducing the risk of fire.

Leather industry

The leather industry has a lengthy history of using enzymes. In the bating process, residual hair and epidermis, together with non-structural proteins and carbohydrates, are removed from the skins, leaving the hide clean, smooth and soft. Traditionally, pancreatic enzymes were employed. Moreover, something in the region of 60% of the input raw materials in leather manufacturing ultimately ends up being discarded and enzyme additions have long been used to help manage this waste. Advances in biotechnology, however, have seen the upsurge in the use of microbially-derived biological catalysts, which are cheaper and easier to produce, for the former applications and the possibility of converting waste products into saleable commodities in the latter.

As well as these improvements on existing uses of biotechnology, new areas of clean application are emerging for tanners. Chemical methods for unhairing hides dissolve the hairs, making for efficient removal, but adding to the treatment cost, and the environmental implications, of the effluents produced, which are of high levels of COD and suspended solids. Combining chemical agents and biological catalysts significantly lessens the process time while reducing the quantities of water and chemicals used. The enzymes also help make intact hair recovery a possibility, opening up the prospect of additional income from a current waste. It has been estimated that, in the UK, for a yearly throughput of 400 000 hides, enzymatic unhairing offers a reduction of around 2% of the total annual running costs (BioWise, 2001). While this may not seem an enormous contribution, two extra factors must be borne in mind. Firstly, the leather industry is very competitive and, secondly, as effluent treatment becomes increasingly more regulated and expensive, the use of clean manufacturing biotechnology will inevitably make that margin greater.

Degreasing procedures are another area where biotechnological advances can benefit both production and the environment, since conventional treatments produce both airborne VOCs and surfactants. The use of enzymes in this role not only gives better results, with a more consistent quality, better final colour and superior dye uptake, but also considerably reduces VOC and surfactant levels. The leather industry is also one of the places where biosensors may have a role to play. With the ability to give almost instantaneous detection of specific contaminants, they may prove of value in giving early warning of potential pollution problems by monitoring production processes as they occur.

Desulphurisation of coal and oil

Microbial desulphurisation of coal and oil represents a further potential example of pollution control by the use of clean technology. The sulphur content of these fossil fuels is of environmental concern principally as a result of its having been implicated in the production of acid rain, since it produces sulphur dioxide (SO_2) on combustion. Most of the work done to date has tended to focus on coal, largely as a result of its widespread use in power stations, though similar worries equally surround the use of high-sulphur oils, particularly as the reserves of low-sulphur fuels dwindle. The sulphurous component of coal typically constitutes between 1 and 5%; the content for oil is much more variable, dependent on its type and original source.

There are two main ways to reduce SO_2 emissions. The first is to lessen the sulphur content of the fuel in the first place, while the second involves removing it from the flue gas. There are a number of conventional methods for achieving the latter, the most commonly encountered being wet scrubbing, though a dry absorbent injection process is under development. At present, the alternative approach of reducing the sulphur present in the initial fuel, works out around five times more expensive than removing the pollutant from the flue gas, though as stock depletion forces higher sulphur coals and oils to be burnt, the economics of this will start to swing the other way. Methods for achieving a sulphur content reduction include washing pulverised coal and the use of fluidised bed technology in the actual combustion itself, to maximise clean burn efficiency.

Sulphur is present in coal in a variety of different forms, both organic and inorganic and biological methods for its removal have been suggested as alternatives to the physical means mentioned above. Aerobic, acidophilic chemolithotrophs like certain of the *Thiobacillus* species, have been studied in relation to the desulphurisation of the inorganic sulphur in coal (Rai, 1985). Microbes of this genus have long been known to oxidise sulphur during the leaching of metals like copper, nickel, zinc and uranium from low grade sulphide ores. Accordingly, one possible application which has been suggested would be the use of a heap-leaching approach to microbial desulphurisation at the mine itself, which is a technique commonly employed for metals. However, though this is, clearly, a cheap and simple solution, in practice it is difficult to maintain optimum conditions for the process. The micro-organisms which have most commonly been used to investigate this possible approach are mesophiles and the rapid temperature increases experienced coupled with the lengthy period of contact time required, at around four to five days form major limiting factors. The use of extreme thermophile microbes, like *Sulfolobus sp*. may offer the way ahead, giving a faster rate of reaction, though demanding the more sophisticated and engineered environment of a bioreactor if they are to achieve their full process efficiency.

The removal of organic sulphur from coal has been investigated by using model organic substrates, most commonly dibenzothiophene (DBT). In laboratory experiments, a number of organisms have been shown to be able to remove organic sulphur, including heterotrophs (Rai and Reyniers, 1988) like *Pseudomonas*, *Rhizobium* and the fungi *Paecilomyces* and chemolithotrophs like *Sulfolobus*, mentioned earlier. These all act aerobically, but there is evidence to suggest that some microbes, like *Desulfovibrio*, can employ an anaerobic route (Holland *et al.*, 1986). While the use of such model substrates has some validity, since thiophenes are the major organic sulphur components in coal, how well their breakdown accurately reflects the situation for the real material remains much less well known. However, this knowledge base is growing, particularly in the light of successful experiments using *Trametes versicolor* ATCC 200801 and *Phanerochaete chrysosporium* ME 446 to desulphurise Tunçbilek lignite, which has a typically high average sulphur content of around 2.59% (Aytar, Sam and Çabuk, 2008).

A range of putative bioreactor designs for desulphurisation have been put forward, involving treatment systems of varying complexity, which may ultimately provide an economic and efficient method for removing sulphur from these fuels prior to burning. However, the state of the art is little advanced beyond the laboratory bench and so the benefit of large scale commercial applications remains to be seen.

Biological Control

The use of insecticides and herbicides, particularly in the context of agricultural usage, has been responsible for a number of instances of pollution and many of the chemicals implicated are highly persistent in the environment. Though there has been a generalised swing away from high dosage chemicals and a widespread reduction in the use of recalcitrant pesticides, worldwide there remains a huge market for this class of agro-chemicals. As a result, this is one of the areas where biotechnological applications may have significant environmental impact, by providing appreciably less damaging methods of pest management. The whole concept of biological control took a severe blow after the widely reported, disastrous outcome of Australia's attempts to use the Cane Toad (*Bufo marinus*) to control the cane beetle. However, in principle, the idea remains sound and considerable research effort has gone into designing biological systems to counter the threat of pests and pathogens. Some of these, in respect of soil-borne plant pathogens and bio-pesticides, are discussed elsewhere in this work and, accordingly, do not warrant lengthy reiteration here.

The essence of the specifically environmental contribution of this type of bio-intervention lies in its ability to obviate the need for the use of polluting chemicals and, consequently, lead to a significant reduction in the resultant instances of contamination of groundwater or land. However, one of the major limitations on the effective use of bio-controls is that these measures tend to act more slowly than direct chemical attacks and this has often restricted their use on commercial crops. In fairness, it must be clearly stated that biotechnology *per se* is not a central, or even necessary, requirement for all of biological control, as many methods rely on whole organism predators, which, obviously, has far more bearing on an understanding of the ecological interactions within the local environment. However, the potential applications of biotechnology to aspects of pest/pathogen/organism dynamics, as examined in other sections of this book, has a supportive role to play in the overall management regime and, thus, there exists an environmental dimension to its general use in this context.

Biological control methods can provide an effective way to mitigate pesticide use and thus the risk represented to the environment and to public health. In addition, unlike most insecticides, bio-controls are often highly target-specific, reducing the danger to other non-pest species. Against this, biological measures typically demand much more intensive management and careful planning than the simple application of chemical agents. Success is much more dependant on a thorough understanding of the life-cycles of the organisms involved and can often be much more of a long-term project. In addition, though high specificity is, generally, a major advantage of bio-control measures, under some circumstances, if exactly the right measure is not put in place, it may also permit certain pests to continue their harmful activities unabated. Considering the huge preponderance of insect species in the world, a large number of which pose a threat to crops or other commodities and thus represent an economic concern, it is small wonder that the global insecticide market has been estimated at over \$10 billion (US) per year. Accordingly, much of the biological control currently in practice relates to this group of animals.

Whole organism approaches

There are three main ways in which whole-organism biological pest control may be brought about. Classical biological control, as with the previously mentioned Cane Toad, requires the importation of natural predators and is principally of use when the pest in question is newly arrived in an area, often from another region or country, having left these normal biological checks behind. Another form of control involves conservation measures aimed at bolstering the predatory species, which may be a valuable approach when natural enemies already exist within the pest's range. However, the third method, augmentation, is more relevant to the concepts of biotechnology and refers to means designed to bring about the increase in effectiveness of natural enemies to a given pest. This may consist simply of artificially rearing them in large numbers for timed release or may extend to more intensive and sophisticated measures like the modification, either by selective breeding or genetic manipulation, of the predator such that it is better able to locate or attack the pest.

One attempt at augmentation which has been tried commercially is the production of entomopathogenic parasitic nematodes. Juvenile stages of the nematodes, which are then only around $500 \, \mu$m long and $20 \, \mu$m wide, can enter soil insects and many carry pathogenic bacteria in their guts. Once ingested, these bacteria pass out of the nematode and multiply inside the insect, typically causing death within a few days. Five species of nematode were originally made available on the US agricultural market, namely, *Steinernema carpocapsae*, *S. riobravis*, *S. feltiae*, *Heterorhabditis bacteriophora* and *H. megidis*, each being effective against different groups of insects. They have subsequently been joined by two more – *S. glaseri* and *H. marelatus*. Despite much research and development effort, the initial results were largely unpredictable, with success against many of the target species, such as wireworms and root maggots, proving elusive. Clearly, farmers and commercial growers will not adopt a biological control which does not reliably offer at least comparable efficacy with traditional insecticides, and consequently for a time uptake was understandably poor. However, technological improvements in organism production, dosage, quality control and product delivery, coupled with increased understanding of optimal timing and greater target application specificity have done much to overcome this. The gap between biological control and conventional agro-chemicals has been significantly narrowed as a result, allowing nematodes to compete effectively in a variety of market sectors.

To date, arguably the most successful application for this technology has arisen not, as many had predicted, in the control of cockroaches, which have been found to be the most vulnerable species to augmented nematode attack (Georgis, 1996), but in controlling garden slugs. Widely available in garden centres and from specialist suppliers, *Phasmarhabditis hermaphrodita* has proven itself to be a highly effective, and entirely specific, method of control on the domestic scale, though here too routine treatment and accurate dosage assessment is the key to success. Augmentation is, obviously, a highly interventionist approach and relies on a regime of continual management to ensure its effectiveness.

There is also a role for the engineered application of biologically derived chemicals in this sector. One example of this is the growing interest in *Azadirachta indica*, the neem, a plant which is found naturally in over 50 countries around the world including India, where its medicinal and agricultural value has been known for centuries. The compound azadirachtin has been identified and isolated from the plant and it has been shown to have broad spectrum insecticidal properties, acting to disturb larval moults and preventing metamorphosis to the imago. Additionally, it also seems to repel many leaf-eating species, and trials involving the direct foliar application of azadirachtin has shown it to be an effective way of protecting crop plants (Georgis, 1996). This duality of action makes it a particularly appealing prospect for wide scale applications, if suitable methods for its production can be made commercially viable.

Semiochemical agents

However, perhaps one of the best examples of the use of such biological technologies in pest control is the development of isolated or synthesised semiochemical agents.

Semiochemicals are natural messenger substances which influence growth, development or behaviour in numerous plant and animal species and include the group known as pheromones, a number of which are responsible for sexual attraction in many insects. This has been successfully applied to control various forms of insect pests, either directly to divert them from crops and trap them, or indirectly to trap their natural enemies in large numbers for introduction into the fields for defence.

For example, crops worldwide suffer severe damage as a result of a number of pentatomid insects, amongst which are several of the common brown stink bugs of North America (*Euschistus* spp.). They arrive late in the growing season and often cause major harm before detection. A major part of bio-control involves obtaining a thorough understanding of their migration patterns and to help achieve it in this case, a pheromone, methyl 2E,4Z-decadienoate, has been produced commercially to aid trapping. The early success of this is being developed to extend its scope in three main directions. Firstly, to capture and eliminate the pests themselves, secondly, to harvest predatory stink bugs for bio-augmentative control programmes and thirdly, to identify more pheromones to widen the number of phytophagous stink bug species which can be countered in this way.

As something of an aside, one interesting and somewhat unusual use has been proposed for this technology. The Siberian moth *Dendrolimus superans* is a vigorous defoliating pest of northern Asian coniferous forests and, though it does not presently occur in North America, its arrival is much feared. In an attempt to provide a first line of defence against this potential threat to native woodlands, it has been suggested that a blend of Z5,E7-dodecadienal and Z5,E7-dodecadienol, which has been shown to act as a powerful sex attractant for male Siberian moths, be deployed at US ports of entry.

However, as illustrated by the case of another pentatomid, *Nezara viridula*, the southern green stink bug, the use of this approach to biological control is not universally applicable. These insects are major agricultural pests affecting a variety of field crops, vegetables, fruits and nuts. While it has been known for some time that sexually mature males produce an attractant pheromone, the active ingredients of which have been identified, early attempts to use this knowledge to exclude them from crops have been of only limited effectiveness. As a result, an alternative method of *Nezara* control has been suggested involving the genetic engineering of its gut symbionts to produce a reduced tolerance of environmental stress. Preliminary work at the Agricultural Research Center, Beltsville, US has isolated and cultured *in vitro* a gram-negative bacterium from the mid-gut of the pest insect, which appears to be a specific symbiont and has been putatively identified as a species of *Yokenella*. This kind of application of transgenic technology may increasingly be the future of biological control for species which do not respond favourably to pheromone trapping.

Not all approaches to bio-control truly qualify as environmental biotechnologies, at least not within the frame of reference used in this book. However, where the use of biological systems results in reduced insecticide use and thus a corresponding lowering of the attendant pollution potential, the net environmental gains of the application of biotechnology are clear.

Bio-Substitutions

The bio-substitution of suitable, less harmful alternatives for many of today's polluting substances or materials is a major potential avenue for the environmentally beneficial application of biotechnology. The question of biofuels and the major renewable contribution which organised, large scale biomass utilisation could make to energy demands is examined in some detail in Chapter 10 and will not, therefore be repeated here. The biological production of polymers, likewise, features in the same section on integrated biotechnology and, though clearly distinctly germane to the present discussion, will also not appear in this consideration. However, the other major use of mineral oils, as lubricants, is an excellent case study of the opportunities, and obstacles, surrounding biotech substitutes. Biodegradable alternatives to traditional lubricating oils have existed for some time, but in many ways they exemplify the pressures which work against novel biological products.

Barriers to uptake

Typically, most of the barriers which they must overcome are non-technical. The pollution of many inland and coastal waters around the world is a well appreciated environmental problem and wider use of these non-toxic, readily biodegradable alternative products could make a huge difference. The main obstacles to wider market acceptance of the current generation of alternative lubricants are neither performance based, nor rooted in industrial conservatism. Cost is a major issue, as bio-lubricants are around twice as expensive as their conventional equivalents, while for some more specialist formulations the difference is significantly greater. Though, inevitably, users need to be convinced of the deliverable commercial benefits, the potential market is enormous. The petrochemical industry has sought to meet the growing demand for more environmentally friendly products by developing biodegradable lubricants based on crude oil. However, with the agricultural sector, particularly throughout Europe, being encouraged to grow non-food crops commercially, there is a clear opportunity for a sizeable vegetable oil industry to develop, though the attitude of heavy industry will prove crucial.

While there is no denying the burgeoning interest in bio-lubricants, the actual machinery to be lubricated is extremely expensive, and enforced down-time can be very costly. Understandably, as a result, few equipment operators are willing to risk trying these new, substitute oils, as original equipment manufacturers (OEMs) are seldom willing to guarantee their performance, not least because vegetable products are often wrongly viewed as inferior to traditional oils.

Biomimetics

The idea of sustainable development is inherent in the science of biomimetics, so it is unsurprising that there are some excellent potential applications of bio-substitution in this field and perhaps one of the best of these is to be found in attempts to defeat aquatic, and especially marine, biofouling. The unwanted accumulation of various forms of life, including algae, microorganisms and sessile animals on submerged structures is a serious economic and practical nuisance to a number of industries. According to US Naval sources, micro-fouling by surface-adhering biofilms can increase drag on a warship by up to 20%, while the presence of macro-fouling by barnacles and other larger organisms can add more than 60% overall (ONR, 2009). This has been calculated as leading to as much as a 40% increase in fuel consumption and up to 10% reduction in a ship's speed. US Navy estimates have put the extra cost that this all entails in supplementary fuel and maintenance at an additional $1 billion per year (ONR, 2009). It is, however, not an issue simply affecting warships. Aside of hulls, any submerged or partially submerged structure, from cables and pipelines to drilling rigs, can suffer biofouling although obviously the relative economic importance it assumes varies.

Conventional anti-fouling treatments rely on the biocidal action of various agents to kill organisms attempting to attach themselves, but they are essentially indiscriminate and tend to both leach and accumulate in the ecosystem, where their toxicity may prove a threat to other non-target species. A common example, tributyltin (TBT), for instance, has been shown to lead to genital abnormalities at concentrations as low as 1 ng/l in the dog whelk *Nucella lapillus* and at 20 ng/l, normal shell formation in the Japanese oyster *Crassostrea gigas* is disrupted. Unsurprisingly, it has been described as the most toxic substance ever introduced into the marine environment (Sonak *et al.*, 2009). In September 2008 the International Convention on the Control of Harmful Anti-Fouling Systems on Ships (AFS Convention) came into force in September 2008, banning the application or re-application of TBT and other organotin compounds, but clearly the need for effective anti-fouling remains. Other chemical based systems have been established to fill the void, and others suggested including, interestingly from a biotechnological standpoint, the possible use of secondary metabolites from cyanobacteria, such as the carboline alkaloid, nostocarboline, isolated from *Nostoc* (Gademann, 2007). As prolific producers of highly chemically diverse secondary metabolites (Smith and Doan, 1999), the cyanobacteria have long been regarded as a potentially valuable source of active anti-fouling agents (Dahms, Ying and Pfeiffer, 2006). However, developments in biomimicry could play a significant role in shaping the future of next-generational anti-fouling products and possibly other surface coatings where effective microbial growth inhibition is also essential.

Unlike marine species such as whales and turtles which are prone to barnacle encrustation, sharks are characteristically untroubled by biofouling and studies of their skin have established that its unique texture coupled with its inherent antimicrobial properties help keep them clean of fouling organisms. Laboratory modelling of shark skin and the subsequent development of a novel biomimetic coating technology – known as Sharklet™ – which features biomimetic topography replicated in polydimethylsiloxane elastomer has yielded a system which has proven remarkably effective at inhibiting bio-fouling. The underlying principle is that both bio-adhesion and surface wet-ability are influenced by surface micro-topography; by finely replicating the features of pelagic shark-skin the biomimetic material has shown a zoospore settlement rate less than 85% that of a conventional smooth surface (Carman *et al.*, 2006). It has also recently been shown to inhibit a range of potentially health threatening microorganisms out of the water, including *Pseudomonas aeruginosa*, Vancomycin-Resistant *Enterococcus* (VRE), *Escherichia coli* and Methicillin-Resistant *Staphylococcus aureus* (MRSA).

Simple bio-substitutions

Not all bio-substitutions need be the result of lengthy chemical or biochemical synthesis or processing and far simpler forms of biological production may provide major environmental benefits. The production of biomass fuels for direct combustion under short rotation coppicing management, described in Chapter 10, is one example. The use of what have been termed 'eco-building materials' formed from hemp, hay, straw and flax and then compressed, as an ecological alternative to conventional materials in the construction industry, is another.

Traditional building approaches have a number of broadly environmental problems. Adequate soundproofing, particularly in home or work settings where traffic, industrial or other noises are a major intrusive nuisance can be difficult or costly to achieve for many standard materials. Walls made from eco-materials have been found to be particularly effective at sound suppression in a variety of applications, including airports, largely due to a combination of the intrinsic natural properties of the raw materials and the compression involved in their fabrication. In a number of trials, principally in Austria where many of the initial materials originated, eco-walls were consistently shown to provide significant improvements in the quality of living and working conditions. Issues of acoustic insulation often go hand-in-hand with the need for thermal insulation and this is one area where simple bio-substitution has made a significant impact. Products such as Thermafleece™, a high density product manufactured from British sheep's wool, offer a thermal conductivity of under 0.04 W/mK and a density of 25 kg/m^3 making them well suited to the purpose. In addition they also exhibit exceptional moisture absorption, with a typical 250 mm layer of insulation being capable of absorbing 2 kg of moisture without any significant detrimental effect on thermal performance. Other natural materials too have become accepted forms of insulation, particularly with the rise of eco-building and the wider drive towards low carbon/low embedded energy products in general within the industry, including flax, hemp and cellulose. As Table 4.5 shows, they compare very favourably

Table 4.5   *Thermal conductivity table*

Material	Conductivity (W/mK)
Cellulose	0.038–0.040
Fibreglass	0.033–0.040
Flax	0.038–0.040
Hemp	0.039–0.043
Mineral wool	0.033–0.047
Sheep's wool	0.037–0.040

in terms of thermal conductivity with the materials traditionally used in lofts or cavity insulation such as fibreglass or mineral wool.

Conventional construction and demolition waste, consisting of concrete rubble, timber fragments, brick shards and the like, poses a considerable disposal problem for the industry, particularly with increasingly stringent environmental regulation and rising storage and landfill costs. Though various recycling initiatives and professional codes of practice have helped ease the situation, there is an obvious advantage in relatively inexpensive, lightweight and sustainable materials which are truly biodegradable. Although uptake has principally been limited to relatively small scale applications, interest has been gradually growing and the momentum seems to be gathering particularly amongst house renovators and the self-build community. The appeal of these and other biological materials for uses in construction, the automotive and aerospace industries is clear, but many of them are still in the quite early stages of their commercial development. Whether they can successfully cross over into mainstream use remains to be seen, but the results to date have certainly been encouraging.

Closing Remarks

As this chapter has shown, pollution and its mitigation have major ramifications in many diverse fields both for industry and in the wider sphere of general human activities. The potential contributions of clean technologies discussed in the final section have enormous bearing on the reduction of contamination *ab initio*, and, clearly, avoiding a problem in the first place is far better than cleaning it up after it has occurred. However, in most cases, current applications of environmental biotechnology to treat pollutants and wastes far outnumber the practical examples of clean bio-manufacturing and so the rest of this book will address this more common use.

References

Aytar, P., Sam, M. and Çabuk, A. (2008) Microbial desulphurization of Turkish lignites by white rot fungi. *Energy and Fuels*, **22** (2), 1196–1199.

BioWise, UK Department of Trade and Industry (2001) Biotechnology Improves Product Quality, Crown copyright.

Brooks, J.R., Barnard, H.R., Coulombe, R. and McDonnell, J.J. (2010) Ecohydrologic separation of water between trees and streams in a Mediterranean climate. *Nature Geoscience*, **3**, 100–104.

Carman, M.L., Estes, T.G., Feinberg, A.W. *et al.* (2006) Engineered antifouling microtopographies – correlating wettability with cell attachment. *Biofouling*, **22** (1), 11–21.

Dahms, H.U., Ying, X. and Pfeiffer, C. (2006) Antifouling potential of cyanobacteria: a mini-review. *Biofouling*, **22** (5), 317–327.

Gademann, K. (2007) Cyanobacterial natural products for the inhibition of biofilm formation and biofouling. *Chimia Natural Products in Drug Discovery*, **61**, 373–377.

Georgis, R. (1996) Present and future prospects of biological insecticides. *Proceedings of the Cornell Community Conference on Biological Control*, April 11–13, Cornell University.

Holland, H., Khan, S., Richards, D. and Riemland, E. (1986) Biotransformation of polycyclic aromatic compounds by fungi. *Xenobiotica*, **16**, 733–741.

Office of Naval Research (2009) Biofouling Prevention Coatings, ONR Program Code 33 and 34.

Rai, C. (1985) Microbial desulfurization of coals in a slurry pipeline reactor using *Thiobacillus ferrooxidans*. *Biotechnology Progress*, **1**, 200–204.

Rai, C. and Reyniers, J. (1988) Microbial desulfurization of coals by organisms of the genus *Pseudomonas*. *Biotechnology Progress*, **4**, 225–230.

Smith, G.D. and Doan, N.T. (1999) Cyanobacterial metabolites with bioactivity against photosynthesis in cyanobacteria, algae and higher plants. *Journal of Applied Phycology*, **11**, 4337–4344.

Sonak, S., Pangam, P., Giriyan, A. and Hawaldar, K. (2009) Implications of the ban on organotins for protection of global coastal and marine ecology. *Journal of Environmental Management*, **90** (1), 96–108.

Case Study 4.1   Mollusc Pollution Detection (Maryland, USA)

Detecting and identifying pollutants is the inevitable first step in any attempt at pollution control and a range of whole organisms and biological systems can help provide sensitive and accurate determinations of contaminants in the environment. A team from the University of the District of Columbia in Washington, DC has successfully shown the efficacy of freshwater clams in this role, demonstrating their ability to not only identify invisible contaminants in water courses, but also help determine their source.

The principle is as simple as it is efficient, which obviously adds to both its cost-effectiveness and ease of use, each of which is a major factor in the uptake of any essentially biological approach to environmental management. As natural

Continued on page 90

—— Continued from page 89 ——

filter-feeders, clams bioaccumulate and biomagnify any water-borne toxins they ingest within their body tissues, effectively becoming a living registry of contaminants within a fairly short time. By placing them in the streams which form tributaries to larger river systems, they produce a biological record of specific water quality, which can later be analysed in the laboratory, allowing an accurate determination of the stream's pollution status to be made. Once this has been achieved, the way is open to begin the process of tracing the pollutants back to their sources, and then ultimately either eradicate or ameliorate them.

This approach has already proven itself in the field, notably detecting a now banned pesticide in a stream in Maryland, which the team believe had been buried some years ago and had subsequently begun slowly leaching into the water course.

The use of appropriate indigenous species to provide a modern day 'canary-in-the-coal-mine' has a clear appeal as a low cost pollution detection system, which lends itself to a wide spectrum of potential applications in developed and developing nations alike.

5

Contaminated Land
and Bio-Remediation

Contaminated land is another example of a widely appreciated, yet often poorly understood, environmental problem, in much the same way as discussed for pollution in the last chapter. That this should be the case is, of course, unsurprising, since the two things are intimately linked, the one being, in essence, simply the manifestation of the other. The importance of land remediation in cleaning up the residual effects of previous human activities on a site lies in two spheres. Firstly, throughout the world, environmental legislation is becoming increasingly stringent and the tightening up of the entire regulatory framework has led to both a real drive for compliance and a much greater awareness of liability issues within industry. Secondly, as the pressure grows to redevelop old, unused or derelict so-called 'brown-field' sites, rather than develop previously untouched 'green-field', the need to remove any legacy of previous occupation is clear. A number of technologies are available to achieve such a clean-up, of which bio-remediation, in its many individual forms, is only one. Though it will, of course, provide the main focus of this discussion, it is important to realise that the arguments presented elsewhere in this book regarding the high degree of specificity which governs technology selection *within* biotechnological applications also applies *between* alternative solutions. In this way, for some instances of contamination, expressly non-biological methods of remediation may be indicated as the best practicable environmental option (BPEO). It is impossible to disassociate contextual factors from wider issues entirely. Accordingly, and to establish the relevancy of the wider setting, alternative remediation techniques will be referred to a little later in this chapter.

The idea of 'contaminated land' is something which is readily understood, yet, like pollution, somewhat more difficult to define absolutely. Implicit is the presence of substances which, when present in sufficient quantity or concentration, are likely to cause harm to the environment or human health. Many kinds of sites may give rise to possible contamination concerns, such as asbestos works, chemical works, garages and service stations, gas works, incinerators, iron and steel works, metal fabrication shops, paper mills, tanneries, textile plants, timber

---

Environmental Biotechnology: Theory and Application, Second Edition Gareth M. Evans and Judith C. Furlong
© 2011 John Wiley & Sons, Ltd.

treatment plants, railway yards and waste disposal sites. This list is not, of course, exhaustive and it has been estimated that in the UK alone something in the region of 360 000 ha (900 000 acres) of land may be affected by contamination in one form or another (BioWise, 2001). Much of this will, of course, be in prime urban locations, and therefore has the potential to command a high market price, once cleaned up.

Since the whole question of contaminated land increasingly forms the basis of law and various professional codes of practice, there is an obvious need for a more codified, legal definition. The version offered in Section 57 of the UK Environment Act, 1995 is a typical example:

> any land which appears ... to be in a condition that ... significant harm is being caused or there is a significant possibility of significant harm ... (or) ... pollution of controlled waters

In this, 'harm' is expressly defined as to 'human health, environment, property'.

As was mentioned earlier, land remediation continues to grow in importance because of pressures on industry and developers. The motive force is, then, a largely commercial one and, consequently, this imposes its own set of conditions and constraints. Much of environmental biotechnology centres on the 'unwanted' aspects of human activity and the clean up of contaminated land is no exception to this general trend. As such, it is motivated by necessity and remedies are normally sought only when and where there is unacceptable risk to human health, the environment and occasionally to other vulnerable targets. In broad terms it is possible to view the driving forces on remediation as characterised by a need to limit present or future liability, increase a site's value, ease the way for a sale or transfer, comply with legislative, licensing or planning requirements or to bolster corporate image or public relations. Generally, one or more of these have to be present before remediation happens.

Having established the need for treatment, the actual remedies to be employed will be based on a realistic set of priorities and will be related to the risk posed. This, of course will require adequate investigation and risk assessment to determine. It is also important to remember in this context that, since the move to remediate is essentially commercial, only land for which remediation is either necessary or worthwhile will tend to be treated and then to a level which either makes it suitable for its intended use or brings it to a condition which no longer poses an unacceptable risk.

It should be apparent, then, from the preceding discussion that the economics of remediation and the effective use of resources are key factors in the whole contaminated land issue. Hence, in purely economic terms, remediation will only take place when one or more of the driving forces becomes sufficiently compelling to make it unavoidable. It will also tend towards the minimum acceptable standard necessary to achieve the required clean-up. This is not an example of industrial self-interest at its worst, but rather the exercise of responsible management, since resources for remediation are typically limited and so their effective use is of great importance. To 'over' remediate any one given site could seriously compromise a company's ability to channel sufficient funds to deal with others. The goal of treating land is to make it suitable for a particular purpose or so that it no longer poses unacceptable risk and once the relevant aim has been achieved, further treatment is typically not a good use of these resources. Generally it would be judged better to devote them to cleaning up other sites, which maximises the potential reuse of former industrial land thereby protecting urban open spaces and the countryside from development pressure. In the long term, the sustainable use of land largely depends on making sure that it is maintained at a level which enables its continued best use for its current or intended purpose. In this respect, discussions of absolute quality become less relevant than a consideration of minimum acceptable standards.

It is also worth noting that the concept of sustainability in relation to the integrated reclamation of brown-field land can vary markedly between developers, site managers and other key players involved in the redevelopment. Such considerations have been shown to have potentially detrimental effects on the successful establishment of 're-greened' brown-field sites (Doick *et al.*, 2009). The recently published Sustainable Remediation Forum (SuRF) UK framework document identified six 'key principles of sustainable remediation', namely:

1. Protection of human health and the wider environment.
2. Safe working practices.
3. Consistent, clear and reproducible evidence-based decision making.
4. Record keeping and transparent reporting.
5. Good governance and stakeholder involvement.
6. Sound science.

SuRF also noted that at times, it may be necessary to make 'non-optimum remediation decisions' in order to maximise the wider benefit of the project (SuRF, 2010). The choice of method and the determination of the final remediation standard will, then, always be chiefly governed by site specific factors including intended use, local conditions and sensitivities, potential risk and available timeframe. For this reason, it is appropriate to take a brief overview of the available technologies at this point, to set the backdrop for the discussions of the specifically biotechnological methods to come.

Remediation Methods

The currently available processes for soil remediation can be divided into five generalised categories:

- Biological,
- Chemical,
- Physical,
- Solidification/vitrification and
- Thermal.

Biological

Biological methods involve the transformation or mineralisation of contaminants to less toxic, more mobile, or less toxic and more mobile, forms. This can include fixation or accumulation in harvestable biomass crops, though this approach is discussed more fully later in Chapter 7.

The main advantages of these methods are their ability to destroy a wide range of organic compounds, their potential benefit to soil structure and fertility and their generally non-toxic, 'green' image. On the other hand, the process end-point can be uncertain and difficult to gauge, the treatment itself may be slow and not all contaminants are conducive to treatment by biological means.

Chemical

Toxic compounds are destroyed, fixed or neutralised by chemical reaction. The principal advantages are that under this approach, the destruction of biologically recalcitrant chemicals is possible and toxic substances can be chemically converted to either more or less biologically available ones, whichever is required. On the downside, it is possible for contaminants to be incompletely treated, the reagents necessary may themselves cause damage to the soil and often there is a need for some form of additional secondary treatment.

Physical

This involves the physical removal of contaminated materials, often by concentration and excavation, for further treatment or disposal. As such, it is not truly remediation, though the net result is still effectively a clean up of the affected site. Landfill tax and escalating costs of special waste disposal have made remediation an increasingly cost-effective option, reversing earlier trends which tended to favour this method. The fact that it is purely physical with no reagent addition may be viewed as an advantage for some applications and the concentration of contaminants significantly reduces the risk of secondary contamination. However, the contaminants are not destroyed, the concentration achieved inevitably requires containment measures and further treatment of some kind is typically required.

Solidification/vitrification

Solidification is the encapsulation of contaminants within a monolithic solid of high structural integrity, with or without associated chemical fixation, when it is then termed 'stabilisation'. Vitrification uses high temperatures to fuse contaminated materials.

One major advantage is that toxic elements and/or compounds which cannot be destroyed, are rendered unavailable to the environment. As a secondary benefit, solidified soils can stabilise sites for future construction work. Never-the-less, the contaminants are not actually destroyed and the soil structure is irrevocably damaged. Moreover, significant amounts of reagents are required and it is generally not suitable for organic contaminants.

Thermal

Contaminants are destroyed by heat treatment, using incineration, gasification, pyrolysis or volatisation processes. Clearly, the principal advantage of this approach is that the contaminants are most effectively destroyed. On the negative side, however, this is achieved at typically very high energy cost, and the approach is unsuitable for most toxic elements, not least because of the strong potential for the generation of new pollutants. In addition, soil organic matter, and, thus, at least some of the soil structure itself, is destroyed.

In situ and *Ex situ* Techniques

A common way in which all forms of remediation are often characterised is as *in situ* or *ex situ* approaches. These represent largely artificial classes, based on no more than where the treatment takes place – on the site or off it – but since the techniques within each do share certain fundamental operational similarities, the classification has some merit. Accordingly, and since the division is widely understood within the industry, these terms will be used within the present discussion.

In situ

The major benefit of approaches which leave the soil where it is for treatment, is the low site disturbance that this represents, which, in many cases, enables existing buildings and features to remain undisturbed. They also avoid many of the potential delays with methods requiring excavation and removal, while additionally reducing the risk of spreading contamination and the likelihood of exposing workers to volatiles. Generally speaking, *in situ* methods are suited to instances where the contamination is widespread throughout, and often at some depth within, a site, and of low to medium concentration. Additionally, since they are relatively slow to act, they are of most use when the available time for treatment is not restricted.

These methods are not, however, without their disadvantages and chief amongst them is the stringent requirement for thorough site investigation and survey, almost invariably demanding a high level of resources by way of both desktop and intrusive methods. In addition, since reaction conditions are not readily controlled, the supposed process 'optimisation' may, in practice, be less than optimum and the true end point may be difficult to determine. Finally, it is inescapable that all site monitoring has an in-built time-lag and is heavily protocol dependent.

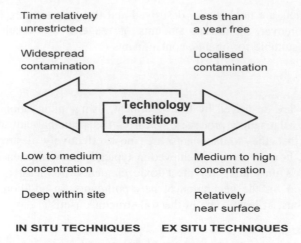

Figure 5.1   *Factors affecting technology suitability*

Ex situ

The main characteristic of *ex situ* methods is that the soil is removed from where it originally lay, for treatment. Strictly speaking this description applies whether the material is taken to another venue for clean-up, or simply to another part of the same site. The main benefits are that the conditions are more readily optimised, process control is easier to maintain and monitoring is more accurate and simpler to achieve. In addition, the introduction of specialist organisms, on those occasions when they may be required, is easier and/or safer and generally these approaches tend to be faster than corresponding *in situ* techniques. They are best suited to instances of relatively localised pollution within a site, typically in 'hot-spots' of medium to relatively high concentration which are fairly near to the surface.

Amongst the main disadvantages are the additional transport costs and the inevitably increased likelihood of spillage, or potential secondary pollution, represented by such movement. Obviously these approaches require a supplementary area of land for treatment and hence they are typically more expensive options.

As Figure 5.1 illustrates, the decision to use *in situ* or *ex situ* techniques is a comparatively straightforward 'black-or-white' issue at the extremes for either option. However, the middle ground between them comprises many more shades of grey, and the ultimate resolution in these cases is, again, largely dependent on individual circumstance.

Intensive and Extensive Technologies

Though the *in situ/ex situ* classification has established historic precedence, of recent times an alternative approach to categorise remediation activities has emerged, which has not yet entirely achieved the same widespread recognition or acceptance, but does, never-the-less offer certain advantages over the earlier approach. Perhaps the most significant of these is that it is a more natural division, based on genuine similarities between technologies in each class. Thus the descriptions 'intensive' and 'extensive' have been suggested.

Intensive technologies can be characterised as sophisticated, fast acting, high intervention strategies, with a heavy demand for resources and high initiation, running and support costs. Their key factors are a fast response and low treatment time, which makes them excellent for heavy contamination conditions, since they can make an immediate lessening in pollutant impact. Soil washing and thermal treatments are good examples of 'intensive' approaches.

Extensive methods are lower-level interventions, typically slower acting, based on simpler technology and less sophisticated engineering, with a smaller resource requirement and lower initiation, running and support costs. These technologies have a slower response and a higher treatment time, but their lower costs make wider application possible, particularly since extensive land remediation treatments do less damage to soil quality. Accordingly, they are well suited to large scale treatment where speed is not of the essence. Examples include composting, the promotion of biological activity *in situ* within the root-zone, precipitation of metal sulphides under anaerobic conditions and the cropping of heavy metal accumulator plants.

All these systems of classification are at best generalisations, and each can be useful at different times, dependent on the purpose of the consideration. They are merely a convenient way of looking at the available techniques and should not be regarded as anything more than a helpful guide. As a final aspect of this, it is possible to examine the various forms of land remediation technologies in terms of their overall functional principle. Hence, the approaches may be categorised as 'destructive', 'separating' or 'containing', dependent on their fundamental mode of operation, as Figure 5.2 illustrates. The principal attraction of this systemisation is that it is defined on the basis of representing the fate of the pollutant, rather than the geographical location of the work or the level of complexity of the technology used, as in the previous cases. In addition, it can also be relatively easily extended to take account of any given technology.

Process Integration

However they are classified, the fact remains that all the individual technologies available each have their limitations. As a result, one potential means of enhancing remediation effectiveness which has become increasingly popular is the use of a combination approach, integrating different processes to provide an overall treatment. The widespread application of this originated some years ago in the USA and the related terms used to describe it, 'bundled technologies' or 'treatment trains' have quickly become commonly used elsewhere. The goal of process integration can be achieved by combining both different fundamental technologies (e.g. biological and chemical) and sequences of *in situ* or *ex situ*,

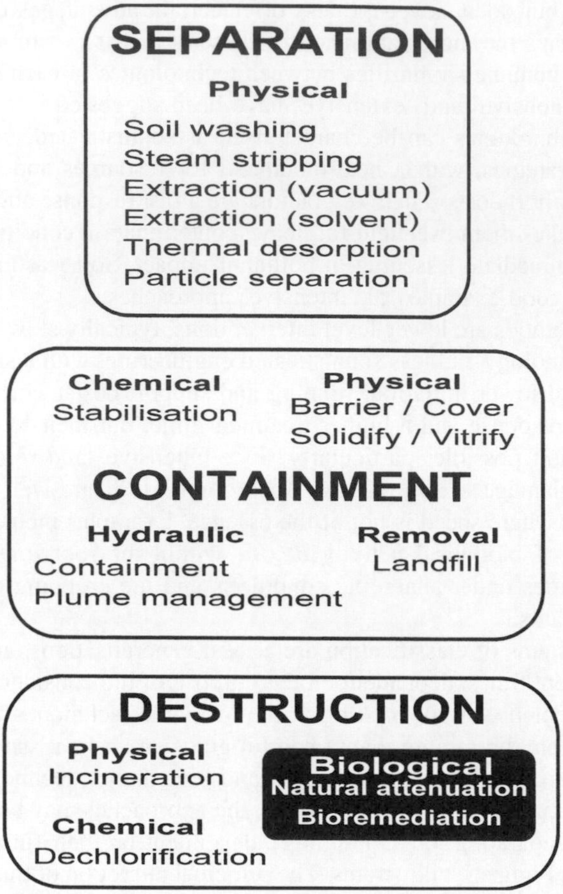

SEPARATION

Physical

Soil washing
Steam stripping
Extraction (vacuum)
Extraction (solvent)
Thermal desorption
Particle separation

Chemical   Physical
Stabilisation   Barrier / Cover
                  Solidify / Vitrify

CONTAINMENT

Hydraulic   Removal
Containment   Landfill
Plume management

DESTRUCTION

Physical   **Biological**
Incineration   Natural attenuation
                  Bioremediation
Chemical
Dechlorification

Figure 5.2   *Technology classification*

intensive or *extensive* regimes of processing. In many respects, such a 'pick-and-mix' attitude makes the whole approach to cleaning up land far more flexible. The enhanced ability this confers for individually responsive interventions stands as one of the key factors in its wider potential uptake. In this way, for example fast response applications can be targeted to bring about a swift initial remediation impact where appropriate, switching over to less engineered or resource-hungry technologies for the long-haul to achieve full and final treatment.

As has been mentioned several times before, commercial applicability lies at the centre of biotechnology, and process integration has clear economic implications beyond its ability simply to increase the range of achievable remediation. One of the most significant of these is that complex contamination scenarios can be treated more cheaply, by the integrated combination of lower cost techniques. This opens up the way for higher cost individual methods to be used only where absolutely necessary, for example in the case of major contamination events or acute pollution incidents. With limited resources typically available for remediation work, treatment trains offer the possibility of maximising their utilisation by enabling responsible management decisions to be made on the basis of meaningful cost/benefit analysis.

This has proven to be an ever more important area over the last 10 years, particularly since increased experience of land remediation successes has removed many of the negative perceptions which were previously commonplace over efficiency, speed of treatment and general acceptability. For many years remediation techniques, and bioremediation especially, were seen in a number of countries as just too costly compared with landfill. As changes in waste legislation in several of these regions have driven up the cost of tipping and wrought considerable restrictions on the amount of biodegradable material entering landfills, the balance has swung the other way, making remediation the cheaper option. There is a certain irony that the very alternative which for so long held back the development of remediation should now provide such a strong reason for its use. In the future, the increasing usage of extensive technologies seem set to continue the trend, since they offer the optimum cost/benefit balance, with intensive processes becoming specialised for fast response or heavy contamination applications. In addition, the 'treatment train' approach, by combining technologies to their maximum efficiency, offers major potential advantages, possibly even permitting applications once thought unfeasible, like diffuse pollution over a large area.

The Suitability of Bioremediation

Bioremediation as a biotechnological intervention for cleaning up the residual effects of previous human activities on a site typically relies on the inherent abilities and characteristics of indigenous bacteria, fungi or plant species. In the present discussion, the emphasis will concentrate on the contribution made by the first two types of organisms. The use of plants, including bioaccumulation, phytoextraction, phytostabilisation and rhizofiltration, all of which are sometimes collectively known as phytoremediation, is examined as part of a separate chapter. Thus, the biological mechanisms underlying the relevant processes are biosorption, demethylation, methylation, metal-organic complexation or chelation, ligand degradation or oxidation. Microbes capable of utilising a variety of carbon sources and degrading a number of typical contaminants, to a greater or lesser extent, are commonly found in soils. By enhancing and optimising conditions for them, they can be encouraged to do what they do naturally, but more swiftly and/or efficiently. This is the basis of the majority of bioremediation and proceeds by means of one of the three following general routes.

Mineralisation, in which the contaminant is taken up by microbe species, used as a food source and metabolised, thereby being removed and destroyed. Incomplete, or staged, decomposition is also possible, resulting in the generation and possible accumulation of intermediate by-products, which may themselves be further treated by other microorganisms.

Cometabolism, in which the contaminant is again taken up by microbes but this time is not used as food, being metabolised alongside the organism's food into a less hazardous chemical. Subsequently, this may in turn be *mineralised* by other microbial species.

Immobilisation, which refers to the removal of contaminants, typically metals, by means of adsorption or bio-accumulation by various micro-organism or plant species.

Unsurprisingly, given the expressly biological systems involved, bioremediation is most suited to organic chemicals, but it can also be effective in the treatment of certain inorganic substances and some unexpected ones at that. Metals and radionuclides are good examples of this. Though, obviously, not directly biodegradable themselves, under certain circumstances their speciation can be changed which may ultimately lead to their becoming either more mobile and accessible or less so. The net result produced in either case can, under the right conditions, be a very effective functional remediation. A list of typical contaminants suitable for bioremediation would include the likes of crude oil and its derivatives, some varieties of fungicides and herbicides, hydrocarbons, glycols, phenols, surfactants and even explosives.

Developments in bio-processing continually redefine the definitive catalogue of what may, and may not, be treated and many chemicals once thought 'impossible' are now routinely dealt with biologically. Table 5.1 reflects the current state of the art, though this is clearly subject to change as new approaches are refined.

As a result, it should be obvious that a large number of opportunities exist for which the application of remediating biotechnologies could have potential relevance. Even so, there are a number of factors which affect their use, which will be considered before moving on to discuss practical treatment issues themselves.

Table 5.1   *Potential for bioremediation of selected contaminants*

Readily possible	Possible under certain circumstances	Currently impossible
Acids	Chlorinated solvents	Asbestos
Alcohols	Cyanides	Asphalt
Aldehydes and ketones	Explosives	Bitumen
Ammonia	PCBs	Inorganic acids
Creosote	PAHs	
Chlorophenols	Pesticides, herbicides	
Crude oil and petroleum	and fungicides	
hydrocarbons	Tars	
Glycols	Timber treatments	
Phenols		
Surfactants		

Factors Affecting the Use of Bioremediation

It is possible to divide these into two broad groups; those which relate to the character of the contamination itself and those which depend on environmental conditions. The former encompass both the chemical nature of the pollutants and the physical state in which they are found in a given incident. Thus, in order for a given substance to be open to bioremediation, clearly it must be both susceptible to, and readily available for, biological decomposition. Generally it must also be dissolved, or at the very least, in contact with soil water and typically of a low–medium toxicity range. The principal environmental factors of significance are temperature, pH and soil type. As was stated previously, bioremediation tends to rely on the natural abilities of indigenous soil organisms and so treatment can occur between 0 and 50 °C, since these temperatures will be tolerated. However, for greatest efficiency, the ideal range is around 20–30 °C, as this tends to optimise enzyme activity. In much the same way, a pH of 6.5–7.5 would be seen as optimum, though ranges of 5.0–9.0 may be acceptable, dependent on the individual species involved. Generally speaking, sands and gravels are the most suitable soil types for bioremediation, with heavy clays and those with a high organic content, like peaty soils, are less well indicated. However, this is not an absolute restriction, particularly since developments in bioremediation techniques have removed the one-time industry maxim that clay soils were impossible to treat biologically.

It should be apparent that these are by no means the only aspects which influence the use of remediation biotechnologies. Dependent on the circumstances, nutrient availability, oxygenation and the presence of other inhibitory contaminants can all play an important role in determining the suitability of bioremediation, but these are more specific to the individual application. A number of general questions are relevant for judging the suitability of biological treatment. The areas of relevance are the likes of the site character, whether it is contained or if the groundwater runs off, what contaminants are present, where they are, in what concentrations and whether they are biodegradable. Other typical considerations would be the required remediation targets and how long is available to achieve them, how much soil requires treatment, what alternative treatment methods are available and at what cost.

Clearly then, there are benefits to the biological approach in terms of sustainability, contaminant removal or destruction and the fact that it is possible to treat large areas with low impact or disturbance. However, it is not without its limitations. For one thing, compared with other technologies, bioremediation is often relatively slower, especially *in situ*, and as has been discussed, it is not equally suitable for all soils. Indeed, soil properties may often be the largest single influence, in practical terms, on the overall functional character of pollution, since they are major factors in modifying the empirical contamination effect. The whole issue may be viewed as hierarchical. The primary influence consists of the contaminants themselves and actual origin of the contamination, which clearly have a major bearing on the overall picture. However, edaphic factors such as the soil type, depth, porosity, texture, moisture content, water-holding capacity, humus content and biological activity may all interact with the primary influences, and/or with each other, to modify the contamination effect, for better or worse. Figure 5.3 is a simplified illustration of this relationship, but as recent work on the ecohydrologic separation of soil water has shown, the inter-functional reality is often vastly more complex (Brooks *et al.*, 2010).

Hence, it is not enough simply to consider these elements in isolation; the functional outcome of the same contaminant may vary markedly, dependent on such site specific differences.

After consideration of the generalised issues of suitability, the decision remains as to which technique is the most appropriate. This is a site-specific issue, for all of the reasons discussed, and must be made on the basis of the edaphic matters mentioned previously, together with proper risk-assessment and site surveys. At the end of all these studies and assessments, the site has been investigated by desk top and practical means, empirical data has been obtained, the resident contamination has been characterised and quantified, its extent determined, relevant risk factors identified and risk assessment has been performed. The next stage is the formulation of a remediation action plan, making use of the data obtained to design a response to the contamination which is appropriate, responsible and safe. At this point, having obtained the clearest possible overview, technology selection forms a major part of this process.

When this has been done, and approval has been gained from the relevant statutory, regulatory or licensing bodies, as applicable, the last phase is to implement the remediation work itself.

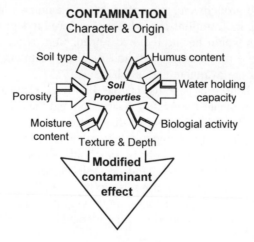

Figure 5.3 *Modification of effect by edaphic factors*

Biotechnology Selection

Although the primary focus of remediation methods commonly falls on technologies dependent on a relatively high engineering component, there is one purely biological treatment option which can be a very effective means of clean-up. Known variously as 'natural attenuation', 'passive remediation', 'bioattenuation' or 'intrinsic remediation', it is appropriate for sites where the contamination does not currently represent a clear danger to human health or the environment. Though it is not an engineered solution, neither is it a 'do nothing' approach as is sometimes stated, since it is not an exercise in ignoring the problem, but a reasoned decision on the basis of the necessary site investigations, to allow nature to take its course. The approach works with natural cycles and the pre-existing indigenous microbial community to bring about the required treatment. The need for a good initial survey and risk assessment is clear, and typically a comprehensive monitoring programme is established to keep a check on progress.

The effectiveness of natural attenuation has been demonstrated by 20 years or more of prior research in the United States, which gave rise to the 'Part 503 Rule'. Issued in February 1993, the *Clean Water Act*, specifically the part of it called *Title 40 of the Code of Federal Regulations, Part 503 – The Standards for the Use or Disposal of Sewage Sludge*, which is commonly referred to as the 'Part 503 Rule' or even simply 'Part 503' set out benchmark limits for the US.

Typical European regulations follow a precautionary limits model, at times referred to as the 'no net gain or degradation' approach, meaning that there should be no overall accumulation of contaminants in the soil, nor any degradation of the soil quality, compared with original levels.

Part 503 is based on risk assessment of selected key pollutants which pose a threat to humans, other animals or plants, making evaluations of a number of different possible pathways, from a direct, 'single incident' scenario, to a lifetime of possible exposure via bioaccumulation. The standard which is set as a result is based on the lowest concentration which was deemed to present an *acceptable* risk.

In this way, higher heavy metal concentrations and cumulative loading rates are permitted than would be allowed under the Europe model, since the ability of soil to lock them up effectively indefinitely has been demonstrated by extensive research. Accordingly US legislation is based on the principle that even if the background level of a given heavy metal species increases over time, its migration or availability for uptake by plants or animals would be precluded by the combined action of the resident microbes and other general soil characteristics. In many ways this has strong echoes of the soil modification of contaminant effect previously discussed.

The engineered solution

If natural attenuation is not appropriate, then some form of engineered response is required, the selection of which will depend on a number of inter-linked factors.

Thus, the type and concentration of the contamination, its scale and extent, the level of risk it poses to human health or the environment, the intended eventual site use, the time available for remediation, available space and resources and any site specific issues all influence this decision. Many of the key issues have already been discussed and the earlier Figure 5.1 set out the factors governing technology transition between the *in situ* and *ex situ* techniques.

Essential Features of Biological Treatment Systems

All biotechnology treatments have certain central similarities, irrespective of the specific details of the technique. The majority of applications make use of indigenous, resident microbes, though in some cases the addition of specialised organisms may be warranted. Thus, the functional biology may be described as a process of bioenhancement or bioaugmentation, or occasionally a mixture of both.

Bioenhancement concentrates solely on the existing micro-fauna, stimulating their activity by the manipulation of local environmental conditions. Bioaugmentation, by contrast, requires the deliberate introduction of selected microbes to bring about the required clean-up. These additions may be unmodified 'wild-type' organisms, a culture selectively acclimatised to the particular conditions to be encountered, or genetically engineered to suit the requirements. Enzyme or other living system extracts may also be used to further facilitate their activity. Some land remediation methods simultaneously bioenhance resident bacteria and bioaugment the process with the addition of fungi to the soil under treatment.

In the final analysis, all biological approaches are expressly designed to optimise the activities of the various micro-organisms (either native to the particular soil or artificially introduced) to bring about the desired remediation. This generally means letting them do what they would naturally do, but enhancing their performance to achieve it more rapidly and/or more efficiently. Effectively it is little different from accelerated natural attenuation and typically involves management of aeration, nutrients and soil moisture, by means of their addition, manipulation or monitoring, dependent on circumstance. However simple this appears, the practical implications should not be underestimated and careful understanding of many inter-related factors is essential to achieve this goal. For example, successful aerobic biodegradation requires an oxygen level of at least 2 mg/l; by contrast, when the major bioremediation mechanism is anaerobic, the presence of any oxygen can be toxic. The presence of certain organic chemicals, heavy metals or cyanides may inhibit biological activity; conversely, under certain circumstances microbial action may itself give rise to undesirable side effects like iron precipitation, or the increased mobilisation of heavy metals within the soil.

In situ techniques

The fundamental basis of *in situ* engineered bioremediation involves introducing oxygen and nutrients to the contaminated area by various methods, all of which ultimately work by modifying conditions within the soil or groundwater. There are three major techniques commonly employed, namely, Biosparging, Bioventing and Injection Recovery. In many respects, these systems represent extreme versions of a fundamentally unified technology, perhaps best viewed individual applications of a treatment spectrum as will, hopefully, become clearer from the descriptions of each which follow.

As set out previously, the major benefits of *in situ* methods are their low intrusion, which enables existing buildings and site features to remain undisturbed, their relative speed of commencement and the reduced risk of contamination spread.

Biosparging

Biosparging is a technique used to remediate contamination at, or below, the water table boundary, a generalised diagram of which appears in Figure 5.4.

In effect, the process involves super-aeration of the groundwater, thereby stimulating accelerated contaminant biodegradation. Though the primary focus of the operation is the saturated zone, the permeability of the overlying soil has a bearing on the process, since increased oxygenation of this stratum inevitably benefits the overall efficiency of remediation.

Air is introduced via pipes sunk down into the contaminated area and forms bubbles in the groundwater. The extra oxygen made available in this way dissolves into the water, also increasing the aeration of the overlying soil, stimulating the activity of resident microbes, which leads to a speeding up of their natural ability to metabolise the polluting substances. The method of delivery can range from relatively simple to the more complicated, dependent on individual requirements. One of the major advantages of this is that the required equipment is fairly standard and readily available, which tends to keep installation costs down. Typically the sparger control system consists of a pressure gauge and relief valve to vent excess air pressure, with associated flow meters and filter systems to clean particulates from the input. More sophisticated versions can also include data loggers, telemetry equipment and remote control systems, to allow for more precise process management. It should be obvious that extensive and comprehensive site investigation, concentrating on site geology and hydrogeology, in particular, is absolutely essential before any work starts.

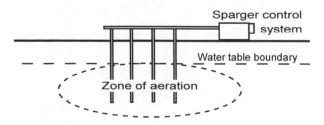

Figure 5.4   *Biosparging*

Bioventing

Bioventing is a technique used to remediate contamination above the water table boundary, and again a generalised diagram appears in the following Figure 5.5. This process also involves super-aeration, again with the intention of stimulating accelerated breakdown of the pollutants present, though this time it is taking place within the soil itself, instead of the groundwater. Bioventing is not generally suitable for remediating sites with a water table within 1 m of the surface, nor for heavy or waterlogged soils, since air flow is compromised under these conditions.

Air is introduced from a compressor pump, via a central pipe, or set of pipes, dependent on the size of the area to be treated, down into the region of contamination. The extra oxygen availability thus achieved, as in the previous approach described, stimulates the resident microbes, which then treat the polluting substances. The air flow through the soil is further driven by vacuum extractors peripheral to the treatment zone, which increases the dissolved oxygen levels of the soil water and thus facilitates uptake by the native micro-organisms. Volatile compounds, which are either present as part of the original contamination, or generated as by-products of the biological treatment, are often mobilised during processing and thus more easily extracted. However, in many practical applications, the air extraction rate is adjusted to maximise decomposition underground, thus reducing a separate requirement for surface treatment of volatile compounds.

As with the biosparger, control devices typically regulate the pressure, filters clean particles from the intake and the flow rate is monitored in operation, with data loggers and telemetry systems again featuring in the more complex applications.

Unsurprisingly, bioventing also requires extensive and comprehensive site investigation before commencement, not least because the proper positioning of the necessary system of pipework is essential to the proper functioning of this technique.

Injection recovery

The injection and recovery method, for which a generalised diagram appears in Figure 5.6, makes use of the movement of ground water through the zone of

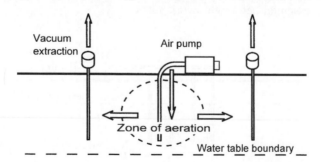

Figure 5.5   *Bioventing*

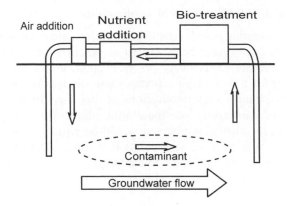

Figure 5.6   *Injection recovery*

contamination to assist the remediation process. Although, as mentioned in the introductory comments, this approach shares many functional similarities with the preceding technologies, it is essentially more sophisticated and refined, with the biological treatment being effectively divided into two complementary stages. Thus, what may be considered a 'virtual' bioreactor is established within the soil matrix, with the actual clean-up activity taking place both within the groundwater and also externally to it.

The major characteristic of this technique is the two well system sunk into the ground, the 'injection well' and the 'recovery well', the former being located 'upstream' of the latter. Nutrients and air are forced down the injection well, and as they flow through the contamination, they stimulate the growth and activity of the indigenous micro-organisms, which begin the process of remediation.

Groundwater, now rich in contaminant, microbes, microbial metabolites and contaminant breakdown products is extracted via the 'recovery well' from beyond the contaminated zone. It then undergoes additional biological treatment above ground in an associated bioreactor vessel, frequently where it is subjected to highly aerobic conditions, before being re-injected, having been further replenished with air and nutrients. This cycle may be repeated many times in the course of treatment. Process control is achieved by having separated out the aeration, nutrient addition and bio-treatment phases into isolated near-episodic events and the facility for direct analysis of the abstracted water enables treatment progress to be monitored with much greater certainty. As a consequence, the injection recovery method neatly overcomes many of the traditional criticisms of *in situ* treatment techniques, particularly in respect of difficulties in ensuring true optimisation of conditions and determining the end point.

Of course, this technique does not avoid the necessity for thorough site investigation and geological surveys, since it is clearly imperative that the particulars of the subterranean water flow, soil depth and underlying geology are known in considerable detail.

Site monitoring for biotechnological applications

Environmental monitoring is well established as a separate science in its own right and many notable books have been written to describe the various approaches and techniques relevant to its many practical applications. It is then, clearly, beyond the scope of this work to reiterate these discussions and the reader is recommended to examine such publications at first hand should detailed information be required. However, since monitoring plays such an important role in the execution of bioremediation plans, it would be wrong not to make mention of some of the more directly relevant points in passing. On particularly relevant aspect in this context is the approach to sampling.

Sampling

Sampling involves two distinct and quite separate activities; firstly, the design of a sampling plan and secondly its practical execution. The first steps are to define the objectives of the programme and obtain agreement from all those concerned as to what will happen to the results produced and what decisions will be taken on the basis of the results obtained. During this process, it is important to consider the scope for any uncertainty which may arise from the intended sampling and analysis and whether the required decisions can be made on the basis of sound data. What analytes or constituents are to be determined and by which analytical methods obviously plays a major part in this, since the accuracy of the intended method must be known if the bioremediation intervention is to be assessed against these measurements.

In respect of the actual sampling methodology, it will be necessary to characterise the samples to be taken adequately in terms of description, location, batch number, size, and so on, as appropriate. The number of incremental samples, and the manner in which they are to be taken is also of great importance to the overall accuracy of the regime and, hence, its value to the wider application of clean-up biotechnology. At the same time, the total volume or mass of the aggregate sample, the size of the final laboratory sample and what steps are taken to reduce the former to the latter must all be carefully decided, as must transport, storage or preparation arrangements.

Safety aspects are also important to take into account at the planning stage, both in terms of the site itself and the expected contamination as well as deciding what specialist equipment or protective clothing will be required. Under certain circumstances, it may be required or advisable to have sampling carried out in the presence of witnesses or representatives of regulatory bodies. This needs to be considered at an early stage and adequate provision made within the plan to accommodate it. Finally, allowance should be made to review the sampling plan, to ensure that it continues to provide adequate data for the purposes required during bioremediation.

The actual analysis and mechanisms of monitoring are well enough covered elsewhere to not warrant consideration here. However, as a final point it is worth noting that for some sites it may be necessary to continue monitoring into the

Figure 5.7   *Illustrative long-term monitoring scheme*

future. Under these circumstances, a comprehensive environmental management and audit scheme can be put in place to monitor environmental effects of such operations and Figure 5.7 shows a suitable illustrative outline.

Under this system, the methodology adopted would be to identify key activities, list the sources of environmental effect to establish quantities and controls and then finally make a determination of their significance. The results would then, of course, feed back into the decision-making process and ultimately help to shape the ongoing environmental management regime of the site.

Ex situ techniques

Again, there are three principal approaches in common use, namely Land Farming, Soil Banking and Soil Slurry Bioreactors. Though inevitably there are distinct similarities between all applications of bioremediation, for obvious reasons of fundamental biology, these techniques are generally more distinct and separate.

The major benefits of *ex situ* methods are the greater ease of process optimisation and control, relatively shorter treatment time and the increased potential for the safe introduction of specialised organisms, if and as required. However, the increased transport costs, additional land requirement and higher levels of engineering often combine to make these technologies more costly options.

Land farming

This technique is effectively accelerated natural attenuation, taking place off-site, within constructed earthwork banking to provide what is essentially a low-tech bioreactor. The pre-treatment stage involves the soil being excavated from site, screened for rocks, rubble and any other oversize inclusions before typically being stored prior to the commencement of actual remediation, either at the original location or on arrival at treatment site.

The processing itself takes place in lined earthworks isolated from the surroundings by an impermeable clay or high density polyethylene (HDPE) liner, as shown diagrammatically in Figure 5.8, and typically relies on the activities

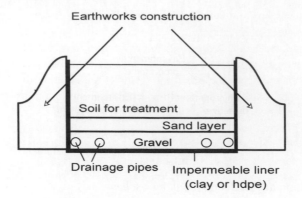

Figure 5.8   *Schematic diagram of land farming*

of indigenous micro-organisms to bring about the remediation, though specialist bacteria or fungi can be added if required. The soil to be treated is laid on a sand layer, which itself stands on a gravel bed, through which a series of drainage pipes have been laid. An impermeable clay or polymer lining isolates the whole system from direct contact with the underlying ground. Water and nutrients are added to stimulate biological activity and soil aeration is maintained by means of turning or ploughing.

The inherent simplicity of the process, however, makes its effectiveness highly dependent on soil characteristics and climatic conditions. For example, heavy clay soils, make attaining adequate oxygenation difficult and uniform nutrient distribution is almost impossible to achieve. In colder climes, it may be necessary to cover the soil to overcome the worst effects of the weather.

Throughout the treatment itself, a process of sampling and monitoring helps to assess progress and compliance with required standards and at completion, the treated soil can be removed either for return to original site or use elsewhere.

Soil banking

In some respects, soil banking is an inverted version of the previous system, ranging from a long row of soil at its simplest, to a more engineered approach, with aeration pipes, a drainage layer, impermeable liner and a reservoir to collect leachate.

Just as with the previous approach, soil is excavated and screened, and is also often stored prior to treatment. As the name suggests, the soil to be processed is formed into banks, sometimes with the addition of filler material like chaff, wood chips or shredded organic matter, if the character of the contaminated soil requires it to improve overall texture, ease of aeration, water-holding capacity or organic matter content. This technique is sometimes termed 'soil composting' because of the similarity it has with the windrow method of treatment for biowaste material, which is described in Chapter 8. It is not a true example of the compost process, though there are a great many functional parallels between these procedures and the same windrow turning equipment may be used in either. Often these rows are covered, either with straw or synthetic blanketing materials, to conserve heat and reduce wash-out. Accordingly, this method is generally better suited to colder and wetter climates and is typically faster than landfarming. Indigenous micro-organisms are again the principal agents of remediation, though specialised bacterial or fungal cultures can be introduced as required, and nutrients added to optimise and enhance their activities.

To further boost the speed and efficiency of this treatment approach, particularly when space is limited, a more sophisticated version, often termed 'engineered biopiling', is sometimes used to ensure greater process control. Leachate is collected in a reservoir and recirculated through the pile to keep the soil moist and return the microbes it contains and a series of pipes within the pile or the underlying drainage layer forces air through the biopile itself. The increased air-flow also permits VOCs to be managed more efficiently and having the whole system above an impermeable geotextile liner prevents leachate migration to the underlying ground.

In both versions of soil banking, a regime of sampling and monitoring is established which again aids process assessment and control. After treatment is concluded, the soil may be returned to the original site for use, or taken elsewhere.

Both landfarming and soil banking are relatively un-sophisticated approaches, effectively utilising the mechanisms of natural attenuation to bring about the necessary clean up, though enhancing and accelerating the process, having first isolated, concentrated and contained the material to be treated. The final commonly encountered technology to be described in this section is a more engineered approach, which works by increasing the levels of water, nutrients and dissolved oxygen available to the microbes.

Soil slurry reactor

In most respects, this system shares essentially similar operating principles to the activated sludge system described in the next chapter, which is used in treating effluents. Figure 5.9 shows a schematic representation of this method.

After excavation, the soil is introduced into a mixing tank, where a slurry is produced by combining it with water. Nutrients are then added to stimulate microbial growth. The suspension formed is transferred to a linked series of well-aerated slurry reactors and micro-organisms within them progressively treat the contaminants. Clarifiers and presses thicken the treated slurry and dewater it, the recovered liquid component being recirculated to the mixing tank to act as the wetting agent for the next incoming batch of soil, while the separated solids are removed for further drying followed by re-use or disposal.

Process selection and integration

However, when complex mixtures of compounds are required to be treated, combining a series of different individual process stages within a series of inter-linked

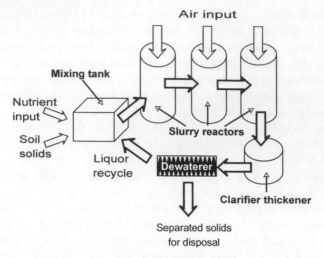

Figure 5.9   *Schematic soil slurry bioreactor system*

bioreactors, may often be a more appropriate and effective response. Dependent on the specific type of contaminants, this may necessitate a sequence of both aerobic and anaerobic procedures, or even one which combines biological and chemical steps to achieve the optimum remediation system. Under such circumstances, clearly each bioreactor features conditions designed to optimise specific biological processes and degrade particular contaminants.

It should be clear from the preceding discussions that the actual process of bioremediation employed will depend on a number of factors, amongst others relating to the site itself, the local area, economic instruments, reasons for remediation and the benefits and limitations of the actual technologies. Hence, it should not be difficult to see that for any given contamination event, there may be more than one possible individual approach and, indeed, as described earlier, the potential will often exist for using integrated combinations of technologies to maximise the effectiveness of the overall response. In this way, though dependent on many external variables, a mix and match assemblage of techniques may represent the individual BPEO. The merging of an *ex situ* treatment, like, for example soil washing via a slurry reactor, to offer an intensive and immediate lessening of pollution effect, with a slower *in situ* process to polish the site to a final level, has much to recommend it, both environmentally and commercially. Accordingly, it seems reasonable to conclude that the prevalence and relative importance of such approaches will continue to grow over the coming years.

Remediation technology selection

As was stated earlier, there are several remediation techniques available, of which bioremediation is just one and, for the most part, regional variables define which approach will tend to be the more commonly used for any given country. This obviously also means that the commercial considerations can, once again, act as a limit on the practical uptake of biotech solutions. Given the typical nature of remediation projects, this inevitably means that the choice between bioremediation and any of the commonly available established 'competitor' technologies, such as containment and encapsulation, excavation for disposal, vacuum extraction, chemical treatment, solvent washing and incineration is not solely based on efficacy or applicability.

Closing Remarks

While there is considerable overlap in technology costs, the economic element remains very largely identifiable in the general trends of use and, of course, biotechnology is not applicable to all forms of pollution. However, global changes in attitude as much as in the escalating costs of special waste disposal have seen the situation for bioremediation improve over the last decade, particularly as the whole concept of sustainable brown-field reclamation has gained momentum.

One major corollary of this has been to see clean-up biotechnologies become recognised as effective and competitive alternatives; it is a trend that would seem to be set to continue in the future.

References

BioWise, UK Department of Trade and Industry (2001) *Contaminated Land Remediation: A Review of Biological Technology*, Crown copyright.

Brooks, J.R., Barnard, H.R., Coulombe, R. and McDonnell, J.J. (2010) Ecohydrologic separation of water between trees and streams in a Mediterranean climate, *Nature Geoscience*, **3**, 100–104.

Doick, K.J., Pediaditi, K., Moffat, A.J. and Hutchings, T.R. (2009) Defining the sustainability objectives of brownfield regeneration to greenspace. *International Journal of Management and Decision Making*, **10**, 282–302.

SuRF, Sustainable Remediation Forum UK (2010) *A Framework for Assessing the Sustainability of Soil and Groundwater Remediation*, Final Report.

Case Study 5.1 Ex-Industrial Site Bioremediation (Utah, USA)

Remediation often plays a significant part in progressing redevelopment projects centred on ex-industrial brownfield sites, with bioremediation enjoying an enviable track-record of successes around the world, especially when the land in question has suffered contamination from petroleum hydrocarbon products.

Never-the-less, applying biotechnological solutions to some of the challenges of hydrocarbon clean-up has not always been uniformly successful across the entire range of potential petroleum contaminants. Depending on the approach used, heavy hydrocarbons in the C15–C40 range, which includes the likes of lubricating oils, drilling mud and crude oil, have at times proven slower or more recalcitrant to treat.

Engaged to remediate a former steel mill facility, Utah-based Terra Nova Biosystems faced C11–C21 diesel range organics (DROs) levels in excess of 100 000 mg/kg, along with broadly similar concentrations of naphthalene and PAH. The site also contained dibenzofurans at 10 000 mg/kg, along with lesser levels of benzene, toluene, ethyl benzene and xylenes.

Treatment involved the use of hydrocarbonoclastic bacteria (HCB) which have the innate ability to degrade hydrocarbon contaminants through their own metabolic action and biosurfactant production. The proprietary process employed required the indigenous HCB population to be artificially augmented and additional oxygen and nutrients supplied to accelerate the *in situ* bioremediation, resulting in effective amelioration within three weeks.

The results far exceeded the US Environmental Protection Agency (EPA)/ Department of Environmental Quality (DEQ) standards for residential soils, with a final total petroleum hydrocarbon gasoline range organic (TPH GRO) concentration of <5 and ~1 mg/kg for all other contaminants.

*Case Study 5.2   Large-Scale Combined Bioremediation
(Köpmanholmens, Sweden)*

Large former industrial sites, particularly those with a long and varied history of use, can often be host to more than one species of contamination, which poses an obvious challenge for remediation.

One solution to this problem lies in deploying a small portfolio of appropriate bioremediation techniques, and by thus combining *in-situ* and *ex-situ* methodologies achieve the necessary synergy for effective treatment. Such an approach self-evidently requires careful management to ensure its efficacy, but, as in the case of the Köpmanholmen's former industrial site, it can prove particularly successful.

The site itself had previously homed chlorine, sulphur and cellulose factories as well as a timber yard, leading to contamination with oil, turpentine, mercury and PAHs. Tasked with the clean-up by the Örnsköldsvik municipality and Naturvårdsverket – the Swedish EPA – MB Envirotech faced the demolition of some 3000 tonnes of concrete foundations, the removal of more than half a kilometre of wastewater pipework and the excavation and *ex-situ* remediation of around 50 000 m^3 of contaminated soil.

The treatment plan they instigated involved a two-phase approach, over an 18–24 months timescale, to first control the extent and spread of contaminants and then subsequently remediate the soil and groundwater to acceptable levels for its future planned use. The biotechnologies involved included bioventing, bioslurping – a combination process to remove free-phase contaminants and simultaneously bioremediate the soils of the vadose zone – and direct groundwater amelioration.

The results of projects such as this, which saw some 15 000 m^3 of soil cleaned within the first six months, has helped to establish bioremediation as a viable option for large-scale applications and raised the bar on the clean-up of multi-use former industrial sites.

6

Aerobes and Effluents

The easiest thing that a sewage works is required to treat is sewage; a large number of industrial or commercial activities produce wastewaters or effluents which contain biodegradable contaminants and typically these are discharged to the sewer. The character of these effluents varies greatly, dependent on the nature of the specific industry involved, both in terms of the likely BOD loading of any organic components and the type of additional contaminants which may also be present. Accordingly, the chemical industry may offer wastewaters with high COD and rich in various toxic compounds, while tannery water provides high BOD with a chromium component and the textile sector is another high BOD effluent producer, with the addition of surfactants, pesticides and dyes. Table 6.1 shows illustrative examples of typical effluent components for various industry sectors.

Table 6.2 shows the illustrative effluent BOD by industry sector, from which it should be apparent that the biodegradable component of any given wastewater is itself highly variable, both in terms of typical values between industries, but also in overall range. Thus, while paper pulping may present an effluent with a BOD of $25\,000\,g/m^3$, sewage returns the lowest of the BODs quoted, clearly underlining the point of this chapter's opening statement.

As might reasonably be expected from the foregoing discussion, the direct human biological contribution to wastewater loading is relatively light. A 65 kg person produces something in the region of 0.1–0.5 kg of faeces per day on a wet weight basis, or between 30 and 60 g of dry solids. The same person also produces around 1–1.5 kg of urine per day, with a total mass of dry solids amounting to some 60–80 g. Of course, the actual effluent arriving at a sewage works for treatment contains the nitrogen, phosphorus and other components originally excreted in the urine or faeces, but in a higher dilution due to flushing water and, often, storm drainage also. Local conditions, climate, details of the sewer system and water availability are, clearly, all potential factors affecting this, though a 49 : 1 ratio of water to solids is fairly typical for developed nations.

Environmental Biotechnology: Theory and Application, Second Edition Gareth M. Evans and Judith C. Furlong
© 2011 John Wiley & Sons, Ltd.

Table 6.1   *Illustrative examples of typical effluent components by industry sector*

Industry sector	Typical effluent component
Chemical industry	High COD, toxic compounds
Distillery	High BOD
Engineering	Oils, metals
Food processing	Fats, starches, high BOD
Paper pulping	Very high BOD, bleaches
Tanning	High BOD, chromium
Textile manufacture	High BOD, surfactants, pesticides, dyestuff
Timber	Preservatives, fungicides

Table 6.2   *Illustrative examples of typical effluent BOD by industry sector*

Industry sector	Typical BOD
Abattoir	2 600
Brewery	550
Distillery	7 000
Landfill leachate	20 000
Paper pulping	25 000
Petroleum refinery	850
Sewage	350
Tannery	2 300

All values in g/m^3.

Sewage Treatment

Looking at sewage works in the strictly literal term, the aims of treatment can be summarised as the reduction of the total biodegradable material present, the removal of any co-existing toxic substances and the removal and/or destruction of pathogens. It is beyond the scope of this book to examine the general, non-biological processes of sewage treatment in great detail, but for the sake of establishing the broader context in which the relevant biotechnology functions, a short description of the main key events follows. It is not, nor is it intended to be, a comprehensive examination of the physical processes involved and the reader is urged to consult relevant texts if this information is authoritatively required.

The typical sewage treatment sequence normally begins with preliminary screening, with mechanical grids to exclude large material which has been carried along with the flow. Paper, rags and the like are shredded by a series of rotating blades known as comminutors and any grit is removed to protect the pumps and ensure free movement of the water through the plant. Primary treatment involves the removal of fine solids by means of settlement and sedimentation, the aim being to remove as much of the suspended organic solid content as possible from the water itself and up to a 50% reduction in solid loading is commonly achieved. At various times, and in many parts of the world, discharge of primary effluent direct to the sea has been permissible, but increasing environmental legislation means that this has now become an increasingly rare option. Throughout the whole procedure of sewage treatment, the effective reduction of nitrogen and phosphorus levels is a major concern, since these nutrients may, in high concentration, lead to eutrophication of the waterways. Primary stages have a removal efficiency of between 5 and 15% in respect of these nutrients, but greater reductions are typically required to meet environmental standards for discharge, thus necessitating the supernatant effluent produced passing to a secondary treatment phase. This contains the main biological aspect of the regime and involves the two essentially linked steps of initial bio-processing and the subsequent removal of solids resulting from this enhanced biotic activity. Oxidation is the fundamental basis of biological sewage treatment and it is most commonly achieved in one of three systems, namely, the percolating filter, activated sludge reactor or, in the warmer regions of the globe, stabilisation ponds. The operational details of the processing differ between these three methods and will be described in more detail later in this section, though the fundamental underlying principle is effectively the same. Aerobic bacteria are encouraged, thriving in the optimised conditions provided, leading to the BOD, nitrogen and ammonia levels within the effluent being significantly reduced. Secondary settlement in large tanks allows the fine floc particles, principally composed of excess microbial biomass, to be removed from the increasingly cleaned water. The effluent offtake from the biological oxidation phase flows slowly upwards through the sedimentation vessels at a rate of no more than 1–2 m per hour, allowing residual suspended solids to settle out as a sludge. The secondary treatment stage routinely achieve nutrient reductions of between 30 and 50%.

In some cases, tertiary treatment is required as an advanced final polishing stage to remove trace organics or to disinfect effluent. This is dictated by water-course requirements, chiefly when the receiving waters are either unable to dilute the secondary effluent sufficiently to achieve the target quality, or are themselves particularly sensitive to some component aspect of the unmodified influx. Tertiary treatment can add significantly to the cost of sewage management, not least because it may involve the use of further sedimentation lagoons or additional processes like filtration, microfiltration, reverse osmosis and the chemical precipitation of specific substances. It seems likely that the ever more stringent discharge standards imposed on waterways will make this increasingly commonplace, particularly if today's concerns over nitrate sensitivity and endocrine disrupters continue to rise in the future.

Process Issues

At the end of the process, the water itself may be suitable for release but, commonly, there can be difficulty in finding suitable outlets for the concentrated sewage sludge produced. Spreading this to land has been one solution which has been successfully applied in some areas, as a useful fertiliser substitute on agricultural or amenity land. Anaerobic digestion, which is described more fully in the context of waste management in Chapter 8, has also been used as a means of sludge treatment. The use of this biotechnology has brought important benefits to the energy balance of many sewage treatment works, since sludge is readily biodegradable under this regime and generates sizable quantities of methane gas, which can be burnt to provide on-site electricity.

At the same time, water resources are coming under increasing pressure, either through natural climatic scarcity in many of the hotter countries of the world, or through increasing industrialisation and consumer demand, or both. This clearly makes the efficient recycling of water from municipal works of considerable importance to both business and domestic users.

Though in many respects the technology base of treatment has moved on, the underlying microbiology has remained fundamentally unchanged and this has major implications in this context. In essence, the biological players and processes involved are little modified from what would be found in nature in any aquatic system which had become effectively overloaded with biodegradable material. In this way, a microcosmic ecological succession is established, with each organism, or group, in turn providing separate, but integrated, steps within the overall treatment process. Hence, heterotrophic bacteria metabolise the organic inclusions within the wastewater, carbon dioxide, ammonia and water being the main by-products of this activity. Inevitably, increased demand leads to an operational decrease in dissolved oxygen availability, which would lead to the establishment of functionally anaerobic conditions in the absence of external artificial aeration, hence the design of typical secondary treatments. Ciliate protozoans feed on the bacterial biomass produced in this way and nitrifying microbes convert ammonia first to nitrites and thence to nitrates, which form the nitrogen-source for algal growth. Though the role of algae in specifically engineered, plant-based monoculture systems set up to reduce the nitrogen component of wastewaters is discussed more fully in the next chapter, it is interesting to note, in passing, their relevance to a 'traditional' effluent treatment system.

One of the inevitable consequences of the functional ecosystem basis underlying sewage treatment plants is their relative inability to cope with toxic chemicals which may often feature in certain kinds of industrial wastewaters. In particular, metabolic poisons, xenobiotics and bactericidal disinfectants may arrive as components of incoming effluents and can prove of considerable challenge to the resident microbes, if arriving in sufficient concentration. This is a fact often borne out in practice; in 2001, considerable disruption was reported as a result of large quantities of agricultural disinfectant entering certain sewage works as a consequence of the UK's foot and mouth disease outbreak. A number of potential consequences arise from such events. The most obvious is that they kill off all or part of the biological systems in the treatment facility. However, dependent on the nature of the substances, in microbially sub-lethal concentrations, they may either become chemically bound to either the biomass or the substrate, or be subject to incomplete biodegradation. The effective outcome of this is that the degree of contaminant removal achievable becomes uncertain and less easily controlled. Partial mineralisation of toxic substances is a particular concern, often leading to the accumulation of intermediate metabolites in the treated wastewater, which may represent the production of a greater biological threat. The incomplete metabolism of these chemicals under aerobic conditions typically results in oxidised intermediary forms which, though less intrinsically toxic than their parent molecules, are often more mobile within the environment. In addition, when the treatment efficiency is subject to monitoring, as intermediate metabolites, these substances may not be picked up by standard analytical techniques, which may result in an unfairly high measure of pollutant removal being obtained.

Moreover, the extension of sewage treatment facilities to ameliorate trade effluents also has implications for the management of true sewage sludge. It is not economically viable to develop processing regimes which do not lead to the concentration of toxic contaminants within the derived sludge. This was shown to be a particular problem for plants using the activated sludge process, which relies on a high aeration rate for pollutant removal, which proceeds by making use of biotransformation, air stripping and adsorption onto the biomass. Adsorption of toxic inorganic substances like heavy metals, or structurally complex organic ones, onto the resident biomass, poses a problem when the microbial excess is removed from the bioreactor, particularly since dewatering activities applied to the extracted sludge can, in addition, catalyse a variety of chemical transformations. Accordingly, sewage sludge disposal will always require careful consideration if the significant levels of these chemicals are not subsequently to cause environmental pollution themselves.

Land Spread

The previous chapter discussed the inherent abilities of certain kinds of soil microbes to remediate a wide range of contaminants, either in an unmodified form, or benefiting from some form of external intervention like optimisation, enhancement or bioaugmentation. Unsurprisingly, some approaches to sewage treatment over the years have sought to make use of this large intrinsic capacity as an un-engineered, low cost response to the management of domestic wastewaters. Thus, treatment by land spread may be defined as the controlled application of sewage to the ground to bring about the required level of processing through the physico-chemical and biological mechanisms within the soil matrix. In most applications of this kind, green plants also play a significant role in the overall treatment process and their contribution to the wider scope of pollutant removal is discussed more fully in the next chapter.

Although it was originally simply intended as a disposal option, in a classic case of moving a problem from one place to another, the modern emphasis is firmly on environmental protection and, ideally, the recycling of the nutrient component. The viability of land treatment depends, however, on the prevention of groundwater quality degradation being afforded a high priority. In the early days of centralised sewage treatment, the effluent was discharged onto land and permitted to flow away, becoming treated over time by the natural microbial inhabitants of the soil. This gave rise to the term a 'sewage farm' which persists today, despite many changes in the intervening years. Clearly, these systems are far less energy intensive than the highly engineered facilities common in areas of greater developed urbanisation.

The most common forms of effluent to be treated by land spread, or the related soil injection approach, are agricultural slurries. According to the European Commission's Directorate General for Environment, farm wastes account for more than 90% of the waste spread on land in Europe and this is predominantly animal manure, while wastes from the food and beverage production industry form the next most important category (European Commission's Directorate General for Environment, 2001). Removal of the constituent nutrients by soil treatment can be very effective, with major reductions being routinely achieved for suspended solids and BOD. Nitrogen removal generally averages around 50% under normal conditions, though this can be significantly increased if specific denitrifying procedures are employed, while a reduction in excess of 75% may be expected for phosphorus. Leaving aside the contribution of plants by nutrient assimilation, which features in the next chapter, the primary mechanisms for pollution abatement are physical filtration, chemical precipitation and microbiological metabolism. The latter forms the focus of this discussion, though it should be clearly understood that the underlying principles discussed in the preceding chapter remain relevant in this context also and will not, therefore be lengthily reiterated here.

The activity is typically concentrated in the upper few centimetres of soil, where the individual numbers of indigenous bacteria and other micro-organisms are huge and the microbial biodiversity is also enormous. This natural species variety within the resident community is fundamental to the soil's ability to biodegrade a wide range of the components in the wastewaters applied to it. However, it must be remembered that the addition of exogenous organic material is itself a potential selective pressure which shapes the subsequent microbial compliment, often bringing about significant alterations as a result. The introduction of biodegradable matter has an effect on the heterotrophic micro-organism population in both qualitative and quantitative terms, since initially there will tend to be a characteristic dying off of sensitive species. However, the additional nutrients made available stimulate growth in those organisms competent to utilise them and, though between influxes, the numerical population will again reduce to a level which can be supported by the food sources naturally available in the environment, over time these microbes will come to dominate the community. In this way, the land spreading of wastewater represents a selective pressure, the ultimate effect of which can be to reduce local species diversity. Soil experiments have shown that, *in extremis*, this can produce a 10-fold drop in fungal species and that *Pseudomonas* species become predominant in the bacterial population (Hardman, McEldowney and Waite, 1994).

With so high a resident microbial biomass, unsurprisingly the availability of oxygen within the soil is a critical factor in the efficiency of treatment, affecting both the rate of degradation and the nature of the end-products thus derived. Oxygen availability is a function of soil porosity and oxygen diffusion can consequently be a rate limiting step under certain circumstances. In general, soils which permit the fast influx of wastewater are also ideal for oxygen transfer, leading to the establishment of highly aerobic conditions, which in turn allow rapid biodegradation to fully oxidised final products. In land that has vegetation cover, even if its presence is incidental to the treatment process, most of the activity takes place within the root zone. Some plants have the ability to pass oxygen derived during photosynthesis directly into this region of the substrate. This capacity to behave as a biological aeration pump is most widely known in relation to certain aquatic macrophytes, notably *Phragmites* reeds, but a similar mechanism appears to function in terrestrial systems also. In this respect, the plants themselves are not directly bio-remediating the input effluent, but acting to bioenhance conditions for the microbes which do bring about the desired treatment.

Septic tank

In many respects, the commonest rural solution to sewage treatment beyond the reach of sewerage, namely, the septic tank, makes use of an intermediate form of land treatment. In the so-called cesspit, a sealed underground tank, collects and stores all the sewage arising from the household. At regular intervals, often around once a month dependent on the capacity, it requires emptying and tankering away, typically for spreading onto, or injection into, agricultural land. By contrast, a septic tank is a less passive system, settling and partially digesting the input sewage, although even with a properly sized and well managed regime the effluent produced still contains about 70% of the original nutrient input. In most designs, this is mitigated by the slow discharge of the liquid via an offtake pipe into a ground soakaway, introducing the residual contaminants into the soil, where natural treatment processes can continue the amelioration of the polluting constituents. There are various types of septic systems in use around the world, though the most common, illustrated in Figure 6.1, is made up of an underground tank, which is linked to some form of *in-situ* soil treatment system, which usually consists of a land drainage of some kind.

Since a system, that is poorly designed, badly installed, poorly managed or improperly sited can cause a wide range of environmental problems, most especially the pollution of both surface and ground waters, their use requires great care. One of the most obvious considerations in this respect is the target soil's ability to accept the effluent adequately for treatment to be a realistic possibility and hence the percolation and hydraulic conductivity of the ground are important factors in the design and long-term success of this method.

Under proper operation, the untreated sewage flows into the septic tank, where the solids separate from the liquids. Surfactants and any fat components tend to float to the top, where they form a scum, while the faecal residues remaining after

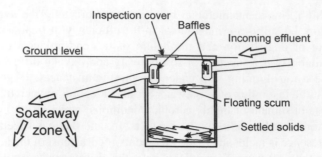

Figure 6.1   *Diagrammatic septic tank*

bacterial action sink to the bottom of the tank, to form a sludge. The biodegradation of the organic effluent in these systems is often only partially complete and so
there tends to be a steady accumulation of sediment within the tank, necessitating
its eventual emptying. This settling effect produces a liquid phase which is permitted to flow out of the tank, along an overflow pipe situated towards the top of the
vessel and is discharged to the soil as previously described. Internal baffles inside
the tank are designed to retain the floating scum layer and prevent undegraded
faeces from leaving the system prematurely. If these biosolids were permitted
to wash out into the soil its ability to treat the septic-tank effluent can readily
become compromised, leading to a reduction in the overall system efficiency.

The drainage arrangements associated with a septic tank system are, arguably,
perhaps the most important part of this whole approach to sewage treatment
and may be considered as effectively forming an underground micro-biological
processing plant. Clearly, it is of vital importance that the soil on any given site
must be suitable for the drainage to function reliably. The only way to be certain
is, of course, by means of a percolation test, though as a general rule, clay soils are
unsuited to this purpose. In circumstances of defined clay strata, particularly when
they exist close to the surface, it is highly unlikely that straightforward drainage
arrangements will prove satisfactory. Even in the absence of a high clay content,
soils which are either too fine or very coarse can also reduce the effectiveness of
this phase of the treatment system. The former can be a problem because, like
clay, it also resists effluent infiltration, the latter because it permits it too quickly
and thus retention time becomes inadequate for the level of treatment needed.

A further consideration which must be addressed in this respect is the position
of the water table, which may cause problems for the drainage system if it
lies within half a metre of the surface. Consequently in areas where this is a
permanent or even seasonal feature, the drains may be established much higher
than would be typical, frequently in close proximity to the soil surface. This
brings its own inevitable set of concerns, not least amongst them being that there
can be a very real possibility of the relatively untreated effluent breaking through
to above ground.

One solution to this potential problem that has been used with some success
is the sewage treatment mound. Formed using clean sand or small gravel, the

mound elevates the system so that it sits a metre or so above the level of the seasonal highest water table. The construction of the mound needs to receive careful consideration to produce a design which suits the local conditions, while also guaranteeing an even distribution of the septic tank effluent throughout the mound. Typically, these systems are intermittently fed by a pump from a collection point and the rate at which the liquid off-take flows through the soil is a critical factor in the correct sizing of the drainage mound. In the final analysis, the sizing of all septic tank systems, irrespective of the details of its specific design, depends on the amount of sewage produced, the type and porosity of soil at the site and the rate at which water flows through it. Proper dimensional design and throughput calculations are of great importance, since the efficacy of septic systems is readily reduced when the set up is overloaded.

Most modern installations use pre-manufactured tanks, typically made of stable polymer and formed in a spherical shape with a short shaft like the neck of a bottle forming a ground level inspection point. They often have a series of internal baffles moulded within them to facilitate the flow of liquids and retention of solids and surface scum, together with the appropriate pipework inlets, outlets and gas-vents. This type of tank has become increasingly popular since they are readily available, easier to site and can be operational much faster than the older concrete designs.

The most common versions of these consisted of two rectangular chambers which were originally built out of brick or stone until the advent of techniques to cast concrete in situ. Sewage digestion was incompletely divided into two stages, with gas venting from the primary chamber and secondary also, in better designed systems. These were sometimes associated with an alternative soil dosing phase, known as seepage pits and soakaways, in which the part-treated effluent arising from the septic tank is discharged into a deep chamber, open to, and contiguous with, the soil at its sides and base. This permitted the free translocation of liquid from the seepage pit into the surrounding soil, the whole of the surrounding ground becoming, in effect, a huge soakaway, allowing dilution and dispersal of the effluent and its concomitant bio-treatment within the body of the soil. In practice, provided the character of the ground is truly suitable for this approach, effluent infiltration and remediation can be very effective. However, if the soil porosity precludes adequate percolation, the potential problems are obvious.

Limits to land application

There are, then, limits to the potential for harnessing the processes of natural attenuation for effluent treatment. While centuries of use across the world testify to the efficacy of the approach for human sewage and animal manures, its application to other effluents is less well indicated and the only truly 'industrial' wastewaters routinely applied to the land in any significant proportion tend to be those arising from food and beverage production. This industry is a consumer of water on a major scale. Dairy production uses between 2 and 6 m^3 of water per 1 m^3 of milk arriving at the plant, the manufacture of preserves requires anything between 10 and 50 m^3 of water per tonne of primary materials consumed and the brewing industry takes 4–15 m^3 of water per tonne of finished beer produced (European Commission's Directorate General for Environment, 2001). A significant proportion of the water is used for washing purposes and thus the industry as a whole produces relatively large volumes of effluent, which though not generally dangerous to human health or the environment, is heavily loaded with organic matter.

The alternative options to land spreading involve either dedicated on-site treatment or export to an existing local sewage treatment works for co-processing with domestic wastewater. The choice between them is, of course, largely dictated by commercial concerns though the decision to install an on-site facility, tanker away to another plant, or land spread is often not solely based on economic factors. Regional agricultural practice also plays an important part, in terms of fertiliser and irrigation requirements as well as with respect to environmental and hydrological considerations. It is of course, a fundamental necessity that the approach selected can adequately cope with both the physical volume of the maximum effluent output on a daily or weekly basis, and the 'strongest' wastewater quality, since each is likely to vary over the year.

Although it is convenient to consider the food and beverage industry as a single group, the effluent produced is extremely variable in composition, depending on the specific nature of the business and the time of the year. However, there are some consistent factors in these effluents, one of which being their typically heavy potassium load. Much of their nutrient component is relatively readily available both for microbial metabolism and plant uptake, which obviously lends itself to rapid utilisation and in addition, the majority of effluents from this sector are comparatively low in heavy metals. Inevitably, these effluents typically contain high levels of organic matter and nitrogen and, consequently, a low C/N ratio, which ensures that they are broken down very rapidly by soil bacteria under even moderately optimised conditions. However, though this is an obvious advantage in terms of their treatability, the concomitant effect of this additional loading on the local micro-biota has already been mentioned. In addition, these effluents may frequently contain heavy sodium and chloride loadings originating from the types of cleaning agents commonly used.

The land application of such liquors requires care since too heavy a dose may lead to damage to the soil structure and an alteration of the osmotic balance. Long term accumulation of these salts within the soil produces a gradual reduction of fertility and ultimately may prove toxic to plants if left to proceed unchecked. Moreover, the characteristically high levels of unstabilised organic material present and the resultant low carbon to nitrogen ratio tends to make these effluents extremely malodorous, which may present its own constraints on available options for its treatment. It is inevitable that issues of social acceptability make land spread impossible in some areas and, accordingly, a number of food and drink manufacturers have opted for anaerobic digestion as an on-site treatment for their process liquors. This biotechnology, which is described in greater detail in Chapter 8, is extremely effective at transforming the organic matter into a methane-rich biogas, with a high calorific value which can be of direct benefit to the operation to offset the heating and electrical energy costs. Under this method, the organic content of the effluent is rapidly and significantly reduced, and a minimum of sludge produced for subsequent disposal.

Nitrogenous Wastes

For those effluents, however, which are consigned to land treatment regimes, the fate of nitrogen is of considerable importance. In aerobic conditions, the biological nitrification processes within the soil produce nitrate from ammonia and organic nitrogen, principally by the chemotrophic bacteria, *Nitrosomonas* and *Nitrobacter*, which respectively derive first nitrites and then finally nitrates. The oxidation of ammonia (NH_3) can be represented as:

$$2NH_3 + 3O_2 \longrightarrow 2NO_2^- + 2H^+ + 2H_2O$$

This reaction releases energy which is subsequently used by *Nitrosomonas* to reduce carbon dioxide. A secondary oxidation of the nitrate produced by Nitrobacter forms nitrate ions, with energy again being released for use by this bacterium.

$$2NO_2^- + O_2 \longrightarrow 2NO_3^-$$

However, in anoxic conditions nitrate compounds can be reduced to nitrogen gas as a result of the activities of various species of facultative and anaerobic soil bacteria, in which the nitrate ion acts as an alternative electron acceptor to oxygen in respiration, as mentioned in Chapter 2. As a result, it becomes possible to view the interlinked processes of nitrogen losses via volatilisation, denitrification and plant uptake as control mechanisms for the nitrogenous component in wastewaters in land applications. Approximately 20–30% of the applied nitrogen is lost in this way, a figure which may rise to as much as 50% under some circumstances, as factors such as high organic content, fine soil particles and water-logging all provide favourable conditions for denitrification within a soil.

Though amelioration processes involving land spreading or injection clearly have beneficial uses for some kinds of wastewaters, in general effluents, particularly those of industrial origin, require more intensive and engineered solutions. In this respect, whether the liquors are treated on-site by the producers themselves, or are tankered to external works is of little significance, since the techniques involved will be much the same irrespective of where they are applied. The contribution of environmental biotechnologies to the safe management of effluents principally centres on microbial action, either in anaerobic digestion where the carbon element is fully reduced, or in aerobic processes which lead to its oxidation. As has been mentioned earlier, the former is covered elsewhere in this book; the rest of this chapter will largely address the latter.

Aeration

Introducing air into liquid wastes is a well established technique to reduce pollutant potential and is often employed as an on-site method to achieve discharge consent levels, or reduce treatment costs, in a variety of industrial settings. It works by stimulating resident biomass with an adequate supply of oxygen, while keeping suspended solids in suspension and helping to mix the effluent to optimise treatment conditions, which also assists in removing the carbon dioxide produced by microbial activity. In addition, aeration can have a flocculant effect, the extent of which depends on the nature of the effluent. The systems used fall into one of two broad categories, on the basis of their operating criteria:

● diffused air systems,
● mechanical aeration.

This classification is a useful way to consider the methods in common use, though it takes account of neither the rate of oxygen transfer, nor the total dissolved oxygen content which is occasionally used as an alternative way to define aeration approaches.

Diffused air systems

The liquid is contained within a vessel of suitable volume, with air being introduced at the bottom, oxygen diffusing out from the bubbles as they rise, thus aerating the effluent.

These systems can be categorised on the basis of their bubble size, with the crudest being coarse open ended pipes and the most sophisticated being specialised fine diffusers. Ultra-fine bubble (UFB) systems maximise the oxygen transfer effect, producing a dense curtain of very small bubbles, which consequently have a large surface area to volume ratio to maximise the diffusion. The UFB system is the most expensive, both to install in the first place and subsequently to run, as it requires comparatively high maintenance and needs a filtered air supply to avoid air-borne particulates blocking the narrow diffuser pores. Illustrative UFB aeration results, based on operational data, obtained from the amelioration of post anaerobic digestion liquor from a horticultural waste processing plant, are shown in Table 6.3.

Though the comparatively simple approaches which produce large to medium sized bubbles are the least efficient, they are commonly encountered in use since they offer a relatively inexpensive solution.

Mechanical aeration systems

In this method, a partly submerged mechanically driven paddle mounted on floats or attached to a gantry vigorously agitates the liquid, drawing air in from the surface and the effluent is aerated as the bubbles swirl in the vortex created.

Table 6.3   *Horticultural waste process liquor analysis before and after 85 day aeration treatment and the associated percentage reductions achieved*

Determinant	Baseline	Post treatment	% reduction
pH	5.8	8.8	–
Conductivity @20 °C	6 950	6320	9.1
BOD total + ATU	15 800	198	98.7
COD	27 200	1990	92.7
Solids particulate 105 °C	6 200	28	99.5
Total dissolved solids	13 700	293	97.9
Ammoniacal nitrogen	515	316	38.6
Total oxidised nitrogen	1.7	0.3	82.4
Kjeldahl nitrogen	926	435	53.0
Nitrite	0.79	0.04	94.9
Nitrate	0.9	0.3	66.7
Sulphate	194	63.4	67.3

All in mg/l except pH (in pH units) and conductivity (in mS/cm).
Results courtesy of Rob Heap, unpublished project report.

Other variants on this theme are brush aerators, which are commonly used to provide both aeration and mixing in the sewage industry and submerged turbine spargers, which introduce air beneath an impeller, which again mixes as it aerates. This latter approach, shown in Figure 6.2, can be considered as a hybrid between mechanical and diffused systems and though, obviously, represents a higher capital cost, provides great operational efficiency. A major factor in this is that the impeller establishes internal currents within the tank. As a result the bubbles injected at the bottom, instead of travelling straight up, follow a typically spiral path, which increases their mean transit time through the body of the liquid and hence, since their residence period is lengthened, the overall efficacy of oxygen diffusion increases.

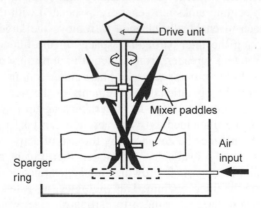

Figure 6.2   *Turbine sparger aeration system*

Table 6.4   *Illustrative oxygen transfer rates for aeration systems at 20°C*

System	Transfer rate (kg O_2/kWh)
Diffused air	
Coarse bubble	0.6–1.2
Medium bubble	1.0–1.6
Fine bubble	1.2–2.0
Brush aerator	1.2–2.4
Turbine sparger aerator	1.2–2.4

The design of the system and the processing vessel is crucial to avoid problems of oxygen transfer, liquid stratification and foaming, all of which can be major problems in operation. The time taken to effect treatment depends on the regime used and the nature of the effluent. In this context, Table 6.4 shows typical oxygen transfer rates for aeration systems at 20 °C.

The value of aeration in the treatment process is not restricted to promoting the biological degradation of organic matter, since the addition of oxygen also plays an important role in removing a number of substances by promoting direct chemical oxidation. This latter route can often help eliminate organic compounds which are resistant to straightforward biological treatments.

Trickling Filters

The trickling or biological filter system involves a bed, which is formed by a layer of filter medium held within a containing tank or vessel, often cast from concrete, and equipped with a rotating dosing device, as shown in a stylised form in Figure 6.3.

The filter is designed to permit good drainage and ventilation and in addition sedimentation and settling tanks are generally associated with the system. Effluent, which has been mechanically cleaned to remove the large particles which might otherwise clog the inter-particulate spaces in the filter bed flows, or is pumped into the rotating spreader, from which it is uniformly distributed across the filter bed. This dosing process can take place either continuously or intermittently, depending on the operational requirements of the treatment works. The wastewater percolates down through the filter, picking up oxygen as it travels over the surface of the filter medium. The aeration can take place naturally by diffusion, or may sometimes be enhanced by the use of active ventilation fans.

The combination of the available nutrients in the effluent and its enhanced oxygenation stimulates microbial growth, and a gelatinous biofilm of microorganisms forms on the filter medium. This biological mass feeds on the organic material in the wastewater converting it to carbon dioxide, water and microbial biomass. Though the resident organisms are in a state of constant growth, ageing and occasional oxygen starvation of those nearest the substrate leads to death of

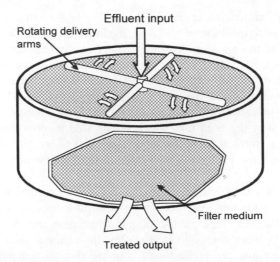

Figure 6.3   *Trickling filter*

some of the attached growth, which loosens and eventually sloughs, passing out of the filter bed as a biological sludge in the water flow and thence on to the next phase of treatment.

The filter media itself is of great importance to the success of these systems and in general the requirements of a good material are that it should be durable and long lasting, resistant to compaction or crushing in use and resistant to frost damage. A number of substances have been used for this purpose including clinker, blast-furnace slag, gravel and crushed rock. A wholly artificial plastic lattice material has also been developed which has proved successful in some applications, but a clinker and slag mix is generally said to give some of the best results. The ideal filter bed must provide adequate depth to guarantee effluent retention time, since this is critical in allowing it to become sufficiently aerated and to ensure adequate contact between the microbes and the wastewater for the desired level of pollutant removal. It should also have a large surface area for biomass attachment, with generous void spaces between the particles to allow the required biomass growth to take place without any risk of this causing clogging. Finally, it should have the type of surface which encourages splashing on dosing, to entrap air and facilitate oxygenation of the bed.

The trickling filters in use at sewage works are squat, typically around 8–10 m across and between 1 and 2 m deep; though these are the most familiar form, other filters of comparatively small footprint but 5–20 m in height are used to treat certain kinds of trade effluents, particularly those of a stronger nature and with a more heavy organic load than domestic wastewater. They are of particular relevance in an industrial setting since they can achieve a very high throughput and residence time, while occupying a relatively small base area of land.

To maximise the treatment efficiency, it is, clearly, essential that the trickling filter is properly sized and matched to the required processing demands. The most important factors in arriving at this are the quality of the effluent itself, its input temperature, the composition of the filter medium, detail of the surface dosing arrangements and the aeration. The wastewater quality has an obvious significance in this respect, since it is this, combined with the eventual clean-up level required, which effectively defines the performance parameters of the system. Although, in an ideal world, the filter would be designed around input character, in cases where industrial effluents are co-treated with domestic wastewater in sewage works, it is the feed rate which is adjusted to provide a dilute liquor of given average strength, since the filters themselves are already in existence. Hence, in practice, the load is often adjusted to the facility, rather than the other way about.

The input temperature has a profound influence on the thermal relations within the filter bed, not least because of the high specific heat capacity of water at $4200 \, J/kg/°C$. This can be of particular relevance in industrial reed bed systems, which are discussed in the following chapter, since a warm liquor can help to overcome the problems of cold weather in temperate climes. By contrast the external air temperature appears to have less importance in this respect. The situation within the reaction space is somewhat complicated by virtue of the non-linear nature of the effect of temperature on contaminant removal. Although the speed of chemical reactions is well known to double for every $10 °C$ rise, at $20 °C$, in-filter biodegradation only represents an increase of 38% over the rate at $10 °C$. Below $10 °C$, the risk of clogging rises significantly, since the activity of certain key members of the microbial community becomes increasingly inhibited.

The general properties of the filter media were discussed earlier. In respect of sizing the system, the porosity and inter-granular spaces govern the interrelation between relative ease of oxygen ingress, wastewater percolation and nutrient to biofilm contact. Clearly, the rougher, pitted or irregular materials tend to offer the greatest surface area per unit volume for microbial attachment and hence, all other things being equal, it follows that the use of such media allows the overall filter dimensions to be smaller. In practice, however, this is seldom a major deciding factor.

In the main, filter systems use rotational dosing systems to ensure a uniform dispersal of the effluent, though nozzles, sprays and mechanised carts are not unknown. The feed must be matched to the medium if the surface aeration effect is to be optimised, but it must also take account of both the fluidity, concentration and quality of the wastewater itself and the character of the resident biofilm.

Since the biological breakdown of effluents within the filter is brought about by aerobic organisms, the effectiveness of aeration is of considerable importance. Often adequate oxygenation is brought about naturally by a combination of the surface effects as the wastewater is delivered to the filter, diffusion from atmosphere through the filter medium and an in-filter photosynthetic contribution from algae. Physical air flow due to natural thermal currents may also enhance the oxygenation as may the use of external fans or pumps which are a feature on some industrial units.

Activated Sludge Systems

This approach was first developed in Manchester, just prior to the outbreak of the First World War, to deal with the stronger effluents which were being produced in increasingly large amounts by the newly emerging chemicals industry and were proving too toxic for the currently available methods of biological processing. Treatment is again achieved by the action of aerobic microbes, but in this method, they form a functional community held in suspension within the effluent itself and are provided with an enhanced supply of oxygen by an integral aeration system. This is a highly biomass-intensive approach and consequently requires less space than filter to achieve same treatment. The main features are shown in Figure 6.4.

The activated sludge process has a higher efficiency than the previously described filter system and is better able to adapt to deal with variability in the wastewater input, both in terms of quantity and concentration. However, very great changes in effluent character will challenge it, since the resident microbial community is generally less heterogeneous than commonly found in filters. Additionally, as a more complex system, initial installation costs are higher and it requires greater maintenance and more energy than a trickling filter of comparable throughput.

In use, the sludge tanks form the central part of a three-part system, comprising a settlement tank, the actively aerated sludge vessels themselves and a final clarifier for secondary sedimentation. The first element of the set-up allows heavy particles to settle at the bottom for removal, while internal baffles or a specifically designed dip pipe off-take excludes floating materials, oil, grease and surfactants.

After this physical pre-treatment phase, the wastewater flows into, and then slowly through, the activated sludge tanks, where air is introduced, providing the enhanced dissolved oxygen levels necessary to support the elevated microbial biomass present. These micro-organisms represent a complex and integrated

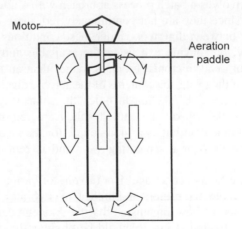

Figure 6.4   *Schematic activated sludge system*

community, with bacteria feeding on the organic content in the effluent, which are themselves consumed by various forms of attached, crawling and free swimming protozoa, with rotifers also aiding proper floc formation by removing dispersed biomass and the smaller particles which form. The action of aeration also creates a circulation current within the liquid which helps to mix the contents of the tank and homogenise the effluent while also keeping the whole sludge in active suspension. Sludge tanks are often arranged in batteries, so that the part-treated effluent travels though a number of aeration zones, becoming progressively cleaned as it goes.

At the end of the central activated phase, the wastewater, which contains a sizeable sludge component by this stage, leaves these tanks and enters the clarifiers. These are often designed so that the effluent enters at their centre and flows out over a series of weirs along the edge of the clarifier. As the wastewater travels outward, the heavier biological mass sinks to the bottom of the clarifier. Typically, collector arms rotate around the bottom of the tank to collect and remove the settled biomass solids which, since it contains growing bacteria that have developed in the aeration tanks, represents a potentially valuable reservoir of process-acclimatised organisms.

Accordingly, some of this collected biomass, termed the Return Activated Sludge (RAS), is returned to the beginning of the aeration phase to inoculate the new input effluent. This brings significant benefits to the speed of processing achieved since otherwise the wastewater would require a longer residence time in which to develop the necessary bacteria and other microbes. It also helps to maintain the high active biomass density which is a fundamental characteristic of this system. The remaining excess sludge is removed for disposal and the clean water flows over another final weir system for discharge, or for tertiary treatment if required.

A similar treatment method sometimes encountered is called aerobic digestion which uses identical vessels to the aeration tanks described, the difference being operational. This involves a batch process approach with a retention period of 30 days or more and since they are not continuously fed, there is no flow through of liquor within or between digesters. Under these conditions, the bacteria grow rapidly to maturity, but having exhausted the available nutrients, then die off leaving a residue of dead microbial biomass, rather than an activated sludge as before. At the end of the cycle, the contents of the aerobic digesters are transferred to gravity thickeners, which function in much the same way as the secondary clarifiers previously described. The settled solids are returned to the aerobic digester not as an innoculant but as a food-source for the next generation, while the clear liquid travels over a separating weir and is returned to the general treatment process.

In effect, then, the 'activated sludge' is a mixture of various micro-organisms, including bacteria, protozoa, rotifers and higher invertebrates forms, and it is by the combined actions of these organisms that the biodegradable material in the incoming effluent is treated. Thus, it should be obvious that to achieve process control, it is important to control the growth of these microbes, which therefore makes some understanding of the microbiology of activated sludge essential. Bacteria account for around 95% of the microbial mass in activated sludge and most of the dispersed growth suspended in the effluent is bacterial, though ideally there should not be much of this present in a properly operating activated sludge process. Generally speaking this tends only to feature in young sludges, typically less than three or four days old, and only before proper flocculation has begun. Ciliates are responsible for much of the removal of dispersed growth and adsorption onto the surface of the floc particles themselves also plays a part in its reduction. Significant amounts of dispersed growth characterises the start-up phase, when high nutrient levels are present and the bacterial population is actively growing. However, the presence of excessive dispersed growth in an older sludge can often indicate that the process of proper floc formation has been interrupted in some way. When floc particles first develop they tend to be small and spherical, largely since young sludges do not contain significant numbers of filamentous organisms and those which are present are not sufficiently elongated to aid in the formation process. Thus, the floc-forming bacteria can only flocculate with each other in order to withstand shearing action, hence the typical globular shape. As the sludge ages, the filamentous microbes begin to elongate, their numbers rise and bacterial flocculation occurs along their length, providing greater resistance to shearing, which in turn favours the floc-forming bacteria. As these thrive and produce quantities of sticky, extra-cellular slime, larger floc particles are formed, the increasingly irregular shape of which is very apparent on microscopic examination of the activated sludge. Mucus secretions from rotifers, which become more numerous as the sludge ages, also contribute to this overall process. Interruption of this formative succession may occur as a result of high toxicity within the input effluent, the lack of adequate ciliated protozoan activity, excessive inter-tank shearing forces or the presence of significant amounts of surfactant.

Process disruption

Toxicity is a particular worry in the operational plant and can often be assessed by microbiological examination of the sludge. A number of key indicators may be observed which would indicate the presence of toxic components within the system, though inevitably this can often only become apparent after the event. Typically, flagellates will increase in a characteristic 'bloom' while higher life forms, particularly ciliates and the rotifers, die off. The particular sensitivity of these microbe species to toxic inputs has been suggested as a potential method of bio-monitoring for toxic stress, but the principle has not yet been developed to a point of practical usefulness.

The floc itself begins to break up as dispersed bacterial growth, characteristic of an immature sludge, returns, often accompanied by foaming within the bioreactor, the progressively reducing growth of microbial biomass leading to a lowered oxygen usage and hence to poor BOD removal. If the toxic event is not so severe as to poison the entire system, as new effluent input washes through the tanks, increasingly diluting the concentration of the contaminating substances and the process recovers, excessive filament formation may occur leading to a condition known as 'filamentous bulking'. As a result, it is sometimes said that toxic inputs favour filamentous bacteria but, with the exception of hydrogen sulphide contamination, this is not strictly true. It is, however, fair to say that the disruption caused by a toxic influx permits their burgeoning growth, particularly since they are generally the fastest group to recover.

By contrast, 'slime bulking' can often occur in industrial activated sludge settings, where the effluent may commonly be deficient in a particular nutrient, most typically either nitrogen or phosphorus. This results in altered floc formation, reduced settling properties and, in some cases, the production of the slimy, greyish foam at the surface of the aeration vessel, which gives this event its name. This greasy, extracellular polymer interferes with the normal settling processes, altering the sludge buoyancy by entrapping air and encouraging foaming. The situation can generally be managed simply by adding appropriate quantities of the missing nutrient, though where relatively easily biodegradable soluble BOD is readily available, it may be necessary to deliberately create higher levels of nitrogen and phosphorus within the system than a straightforward analysis might otherwise indicate.

Foaming can be a significant and unsightly nuisance in operational facilities and, as has been discussed, may occur as a result of either nutrient deficiency or the growth of specific foam-generating filamentous organisms. Microscopic examination of the fresh foam is often the best way to determine which, and thus what remedial action is necessary.

Typical protozoans present in the sludge include amoebae, ciliates and flagellates and, together with rotifers, they play secondary roles in the activated sludge treatment of wastewaters. The presence or absence of particular types can be used as valuable biological indicators of effluent quality or plant performance. In this way, the incidence of large numbers of amoeba often suggests that a shock loading has taken place, making large quantities of food available within the system, or that the dissolved oxygen levels in the tanks have fallen, since they are better able to tolerate conditions of low aeration. A large flagellate population, particularly in mature sludges, suggests the persistence of appreciable quantities of available organic nutrients, since their numbers are usually limited by competition with bacteria for the same dissolved foodstuff. Since ciliates, like rotifers, feed on bacteria, their presence indicates a healthy sludge, as they typically blossom after the floc has been formed and when most of the effluent's soluble nutrients have been removed. As protozoa are more sensitive to pH than floc-forming bacteria, with a typical optimum range of 7.0–7.4 and tolerating 6.0–8.0, they can also provide a broad measure of this parameter in the system.

The population of rotifers seldom approaches large numbers in activated sludge processes, though they never-the-less perform an important function. Their principal role is the removal of dispersed bacteria, thus contributing to both the proper development of floc and the reduction of wastewater turbidity. Taking the longest time of all members of the microbial community to become established in the sludge, their presence indicates increasing stabilisation of the organic components of the effluent.

Organic loadings

Calculating the organic loadings for a given activated sludge system is an important aspect of process control. Measuring the BOD of the incoming wastewater gives a value for the amount of biodegradable matter available for microbial use, which can be used together with an estimate of the resident biomass to derive a relationship termed the food to micro-organism (F/M) ratio. This, which is also sometimes known as the organic loading rate, is given as follows:

$$F/M = \frac{\text{mass of BOD applied to the biological phase each day}}{\text{total microbial biomass in the biological phase}}$$

The F/M ratio is a useful indication of anticipated micro-organism growth and condition, a high F/M value yielding rapid biomass increase, while a low one suggests little available nutrients and consequently slow growth results. Clearly, the total active biomass content in an activated sludge system, which is termed the Mixed Liquor Suspended Solids (MLSSs), is an important factor in process efficacy. Accordingly, it is routinely measured at sewage works being important in the calculation of the F/M ratio, which can be more properly defined as:

$$F/M = \frac{\text{flow rate } (m^3/d) \times \text{BOD } (kg/m^3)}{\text{volume of sludge tank } (m^3) \times \text{MLSS } (kg/m^3)}$$

Although the preceding systems are the most common forms likely to be seen in use, a number of other systems exist which may sometimes be encountered, some of which will be briefly outlined for the sake of completeness.

Deep Shaft Process

In many respects this is an activated sludge derivative, which was born out of ICI's work on the production of proteins from methanol in the 1970s. Figure 6.5 shows the main features of the system, which is based around a shaft 50–100 m deep.

The shaft contains the wastewater to be treated, compressed air being blown in at the base, which travels up the central section, setting up an opposing counter flow in the outer part of the shaft. Screened secondary effluent is allowed to settle and a portion of the sludge produced is returned to the input zone, just as in a traditional activated sludge tank, though degassing is required to remove nitrogen and carbon dioxide bubbles from the floc to allow for proper sedimentation.

The high pressures at the base forces far more oxygen into the solution than normal, which aids aeration enormously and allows the process to achieve an

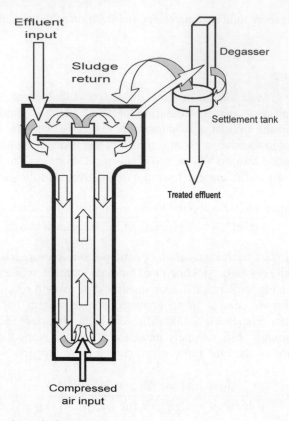

Figure 6.5   *ICI deep shaft process*

oxygen utilisation of around 90%, which is some 4.5 times better than conventional activated sludge systems. The bubble contact time produced, averaging 90 seconds or more, is over six times longer than in standard diffused air systems. It has a low footprint, making it ideal for use in restricted areas.

Pure Oxygen Systems

With process efficacy so closely dependent on aeration and the ability to support a high microbial biomass, the use of pure oxygen to enhance the effective levels of the gas dissolved in the effluent has an obvious appeal. The UNOX process, which was developed by the Union Carbide Corporation is probably amongst the best known of the pure oxygen activated sludge systems and Figure 6.6 shows the general layout of the bioreactor vessels.

Pure oxygen obviously gives a better oxygen transfer rate per unit volume of the bioreactor than can be achieved using conventional aeration methods. In turn, this allows a heavier organic loading per unit volume to be treated

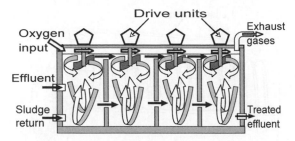

Figure 6.6   *The UNOX™ pure oxygen system*

compared with ordinary air-fed systems, which enables this system to be used to deal with stronger effluents and permits a high throughput where space is restricted. Typically these systems are fed using liquid oxygen tanks.

Despite their clear advantages, pure oxygen systems suffer with some major drawbacks. For one thing, the capital costs involved in installing them in the first place are considerable, as are their running costs and maintenance requirement. The pure oxygen itself represents an explosion risk, thus necessitating intrinsically safe operational procedures and, in addition, leads to accelerated corrosion of the equipment used. However, for some applications and for certain kinds of effluents, they can prove particularly appropriate.

The Oxidation Ditch

This is sometimes used for sewage treatment and is characterised by a constructed ellipsoidal ditch, in which the effluent is forced to circulate around the channel by brush aerators. The ditch itself is trapezoidal in cross section to maintain uniform effluent velocity throughout the channel. Effluent is fed into the system without any prior primary sedimentation and typically gives rise to only 50% of the surplus sludge produced by a typical activated sludge process.

The Rotating Biological Contactor

This system, shown in Figure 6.7, is a derivative of the biological filter. It effectively combines the advantages of this previously described approach, like the absence of a complicated settlement system for sludge return and a low maintenance requirement with the smaller footprint and long microbial exposure characteristic of the active sludge process.

They have submerged internal disc baffles which act as sites for the attached growth of biomass, which are slowly turned by electric motor causing the microbes to be alternately aerated and immersed in the effluent. Rotating biological contactors are typically used for small installations, and are particularly useful for applications with high seasonal variations like caravan site sewage systems.

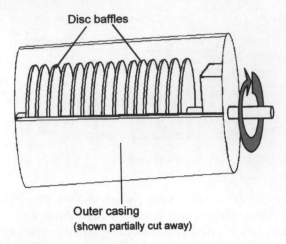

Figure 6.7   *Rotating biological contactor*

Membrane Bioreactors

This system, instead of utilising conventional methods of gravity settlement, achieves the desired biomass retention by means of a cross-flow filtration process, as shown in Figure 6.8.

The development of effective methods of micro- and ultra-filtration has opened up the potential for using membrane bioreactor technology on various forms of domestic and industrial effluents. There are three general types of reactor systems which have been developed, namely, solid/liquid separation, gas permeable and extractive systems. The membrane element allows the passage of small molecules, but retains the total resident microbial population. As a result, the cumulative overall bio-activity and the resultant speed of remediation is boosted, since not only are micro-organisms no longer lost with wash-out flow, but also, conditions for even the slowest-growing member species of the microbial community are able to be adequately enhanced. This is of particular relevance to xenobiotics and the more recalcitrant components of wastewaters, as their biological breakdown is often brought about by bacteria which themselves have a relatively long establishment period within the population. The high biomass levels within the bioreactor itself obviously necessitate abundant readily available oxygen, though the high organic loading and efficient intra-system microbial conservation combine to make the hydraulic retention time entirely independent of the solids themselves. Since the membrane allows gaseous transport while constraining the biological phase, there is provision within the reactor for bubble-less aeration and oxygenation consequently can take place over a relatively large surface area, thereby improving the efficiency of this process. In addition, the membrane itself may become an attachment zone for biofilm formation.

Thus, the membrane bioreactor can offer a greater degradation capacity for persistent chemicals, making possible the biological removal of benzene,

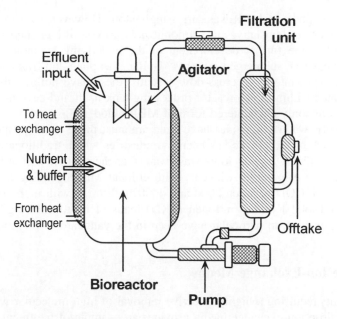

Figure 6.8   *Schematic membrane bioreactor*

nitrobenzene, dichloroaniline and polyaromatic hydrocarbons (PAHs), for example which represent a significant risk, both to the environment and human health, due to their high toxicity. Removal efficiency for these substances can approach 99%.

The membrane bioreactor has proved its suitability as an efficient system for degradation of recalcitrant compounds and significantly higher biomass concentrations and utilisation rates are routinely achieved than in corresponding alternative treatment systems. In common with most operational, rather than experimental, biological detoxification processes, not all of the contaminants present in the effluent are typically completely converted into carbon dioxide and water, a certain percentage being turned into metabolic by-products instead, though this can amount to less than 5% in a well managed bioreactor system. Part of this involves the gradual and controlled introduction of novel wastewater elements, to ensure that acclimatisation is maximised and any potential tendency for 'shock loadings' avoided. This is a clear example of the value of permitting optimised microbial adaptation to the individual application.

These systems are, of course, more expensive than the conventional activated sludge or trickling filters, but produce a much smaller quantity of excess sludge for subsequent disposal of treatment. In addition, they produce an elevated COD removal and would seem particularly well suited to use in small scale plants where the production of high quality final effluent is a priority. Never-the-less, for most of the first decade of its commercial history, and despite this evident promise, the uptake of MBR was surprisingly slow. However, over recent years, the combination of legislative and economic pressures has driven increasing interest and led to the further development of the underlying technology itself to produce the latest submerged systems. Unlike the plate-and-frame or tubular membranes encountered in cross-flow systems, these have hollow fibres or flat sheets located within the bioreactor processing tank itself and have now become the most commonly encountered form of MBR (Judd, 2006).

Predictably, water companies have been amongst those to adopt the technology, but much of the uptake has been by industries which traditionally produce particularly difficult or recalcitrant wastewaters, as the regulation of effluent quality has become ever more stringent. With effluent MLSS ranging from 10 000 to 20 000 mg/l, COD between 2000 and 4000 mg/l, often with high ammoniacal nitrogen and total Kjeldahl nitrogen (TKN) loads of up to 1000 mg/l, pharmaceutical companies, in particular have been in the vanguard of this move.

Cellulose Ion-Exchange Media

For effluents requiring a highly selective removal of high molecular weight proteins, cellulose ion-exchange media provides an example of treatment involving the use of isolated biologically derived materials. The ion exchange medium is replenished with brine as required, and the proteins collected are removed in the resulting saline solution, for subsequent coagulation and drying. This enables a valuable material to be recovered, typically for use as an animal foodstuff, while reducing the wastewater BOD by 90% or more.

Sludge Disposal

Many of the treatment processes described in this chapter give rise to primary or secondary sludges. Typically, these by-products require disposal and, like many forms of solid waste, a proportion have been consigned to either landfill or incineration. For some treated sludges, especially those derived from domestic sewage or food residuals, agricultural use has been an option, often requiring additional treatments to ensure its freedom from human pathogens, before land spreading or injection beneath the surface. The effectiveness of microbes in metal sequestration means, inevitably, that most treated sludges have a degree of heavy metal contamination, which itself makes possible the accumulation of these contaminants in soils exposed to these products. In addition, there are increasingly stringent controls on the release of nitrogen to the environment, particularly within escalating European Union legislation regarding nitrogen vulnerable zones. It would seem, then, that the future land use of 'spent' sludges is likely to be somewhat more heavily regulated than previously.

Closing Remarks

In many respects, the treatment of effluents by biological means is of particular importance to any consideration of environmental biotechnology, since it represents the central point of the previously mentioned intervention triangle, having simultaneous relevance to manufacturing, waste management and pollution control. The majority of manufacturing companies produce wastewaters that contain organic contaminants of one form of another and the traditional route previously in common use, discharge to sewer, watercourse or the sea, is becoming less attractive due to environmental legislation and rising disposal costs. As a result, for an increasing number of companies there is a growing requirement to treat their own effluents and biotechnological processes can often prove the most cost-effective means to achieve this goal.

References

European Commission's Directorate General for Environment (2001) *Survey of Wastes Spread on Land – Final Report*, Office for Official Publications of the European Communities, Luxembourg, pp. 2 and 55.

Hardman, D., McEldowney, S. and Waite, S. (1994) *Pollution: Ecology and Biotreatment*, Longman, Essex, pp. 81–82.

Judd, S. (2006) *The MBR Book; Principles and Applications of Membrane Bioreactors in Water and Wastewater Treatment*, Elsevier, Oxford, pp. 134–149.

Case Study 6.1   'CSI Seaside' (England and Wales)

Although around 97% of the beaches in England and Wales successfully met the necessary European quality standards in 2008, a small proportion struggle with persistent pollution problems, especially after incidents of heavy rainfall.

In an attempt to improve monitoring and ultimately reduce the incidence of contamination, in 2009 the Environment Agency extended state-of-the-art forensic DNA techniques to the large-scale analysis of bathing waters. In what is said to be the first initiative of its kind in the world, the 'CSI Seaside' project – implemented at a number of sites around the coastline – brings microbial source tracking (MST) to the problem of identifying the origin of diffuse faecal matter in the water.

The basic principle behind MST involves screening laboratory-isolated DNA preparations from water for the presence of particular *Bacteroidetes* species which are naturally either only human-specific or animal-specific. Genotype determination of *Bacteroidetes* in this way provides an unambiguous indicator of the original host animal and thus makes it possible to determine the fundamental nature of the faecal contamination, enabling sewage to be quickly discriminated from agricultural animal

Continued on page 144

__ Continued from page 143 _____

manures. The method can also be adapted to be quantitative rather than simply indicative, allowing data to be gathered on the relative preponderance of marker bacteria from each host group in the total population of faecal bacteria present.

Tackling diffuse pollution is invariably a more difficult, more technically demanding and typically more costly challenge than addressing point source contamination. Identifying the original character of the faecal pollution obviously establishes the likely type of source, simplifying the job of locating the specific agricultural run-off or sewage discharge responsible on the ground and reducing the delay before remedial action can be implemented.

Case Study 6.2   Hydrogen from Wastewater (California, USA)

Effluents are generally viewed as problems in need of treatment and ultimately safe disposal, rather than being seen as potentially useful resources in their own right, but the growing success of a new approach, microbial electrolysis (ME), could change that. Although ME – a process for producing hydrogen from wastewater – was only invented in 2003, it has developed rapidly and now offers yields that approach 100%.

The Napa Wine Company in Oakville California recently began hosting a pioneering continuous-flow ME demonstration plant, developed by Professor Bruce Logan of the Department of Environmental Engineering at Penn State University, able to process around 1000 l/day of effluent. ME cells comprise two electrodes immersed in a liquid – in this case winery wastewater, which arises from the wine making process itself and a range of allied activities including washing down and equipment cleaning. Naturally occurring electrochemically active bacteria convert the oxidisable material in the effluent to CO_2, protons and electrons. The electrons transfer to the anode, flowing to the cathode along an electric circuit, while the protons (H^+ ions) also migrate towards the cathode by diffusing through a cation membrane, uniting with the electrons to form hydrogen gas. To run, the reaction requires a small addition of external electricity, typically a voltage of less than 0.2 V, and provides a relatively high yield of hydrogen.

The part-treated water remaining after electrolysis enters the site's pre-existing effluent treatment facilities before being recycled for irrigation purposes, while some of the gas produced will be used on-site in a hydrogen fuel cell, establishing the potential to develop ME into a source of clean energy in the future.

7

Phytotechnology and Photosynthesis

From a practical standpoint, phytotechnology is the use of plants in environmental biotechnology applications, and draws on many of the characteristics which have already been described. In this respect, it does not represent a single unified technology, or even application, but rather is a wider topic, defined solely by the effector organisms used. Thus the fundamental scope of this chapter is broader and the uses and mechanisms described somewhat more varied than for many of the preceding biotechnologies discussed.

Plants of one kind or another can be instrumental in the biological treatment of a large number of substances which present many different types of environmental challenges. Accordingly, they may be used to remediate industrial pollution, treat effluents and waste waters or solve problems of poor drainage or noise nuisance. The processes of bioaccumulation, phytoextraction, phytostabilisation and rhizofiltration are collectively often referred to as phytoremediation. Although it is sometimes useful to consider them separately, in most functional respects, they are all aspects of the same fundamental plant processes and hence there is much merit in viewing them as parts of a cohesive whole, rather than as distinctly different technologies. It is important to be aware of this, particularly when reading a variety of other published accounts, as the inevitable similarities between descriptions can sometimes lead to confusion. Moreover, the role of phyto*technology* is not limited solely to phyto*remediation* and this discussion, as explained above, is more deliberately inclusive of wider plant-based activities and uses.

Despite the broad spectrum of potential action exhibited by plants in this respect, there are really only three basic mechanisms by which they achieve the purpose desired. In essence, all phytotechnology centres on the removal and accumulation of unwanted substances within the plant tissues themselves, their removal and subsequent volatilisation to atmosphere or the facilitation of in-soil treatment. Plant-based treatments make use of natural cycles within the plant and its environment and, clearly, to be effective, the right plant must be chosen. Inevitably, the species selected must be appropriate for the climate, and it must, obviously, be able to survive in contact with the contamination to be able to

Environmental Biotechnology: Theory and Application, Second Edition Gareth M. Evans and Judith C. Furlong
© 2011 John Wiley & Sons, Ltd.

accomplish its goal. It may also have a need to be able to encourage localised microbial growth.

One of the major advantages of phytotechnological interventions is their almost universal approval from public and customer alike and a big part of the appeal lies in the aesthetics. Healthy plants, often with flowers, make the site look more attractive, and helps the whole project be much more readily accepted by people who live or work nearby. However, the single biggest factor in its favour is that plant-based processes are frequently considerably cheaper than rival systems, so much so that sometimes they are the only economically possible method. Phytoremediation is a particularly good example of this, especially when substantial areas of land are involved. The costs involved in cleaning up physically large contamination can be enormous and for land on which the pollution is suitable and accessible for phyto-treatment, the savings can be very great. Part of the reason for this is that planting, sowing and harvesting the relevant plants requires little more advanced technology or specialised equipment than is readily at the disposal of the average farmer.

The varied nature of phytotechnology, as has already been outlined, makes any attempt at formalisation inherently artificial. However, for the purposes of this discussion, the topic will be considered in two general sections, purely on the basis of whether the applications themselves represent largely aquatic or terrestrial systems. The reader is urged to bear in mind that this is merely a convenience and should be accorded no particular additional importance beyond that.

Terrestrial Phyto-Systems (TPS)

The importance of pollution, contaminated land and the increasing relevance of bioremediation have been discussed in previous chapters. Phytoremediation methods offer significant potential for certain applications and, additionally, permit much larger sites to be restored than would generally be possible using more traditional remediation technologies. The processes of photosynthesis described earlier in this chapter are fundamental in driving what is effectively a solar-energy driven, passive and un-engineered system and hence may be said to contribute directly to the low cost of the approach.

A large range of species from different plant groups can be used, ranging from pteridophyte ferns, to angiosperms like sunflowers, and poplar trees, which employ a number of mechanisms to remove pollutants. There are over 400 different species considered suitable for use as phytoremediators. Amongst these, some hyper-accumulate contaminants within the plant biomass itself, which can subsequently be harvested, others act as pumps or siphons, removing contaminants from the soil before venting them into the atmosphere, while others enable the biodegradation of relatively large organic molecules, like hydrocarbons derived from crude oil. However, the technology is relatively new, only recently emerging from the development phase. The first steps toward practical bioremediation using various plant-based methods really began with research in the early 1990s, with a number of the resulting techniques having been used subsequently in the field to good effect.

In effect, phytoremediation may be defined as the direct *in situ* use of living green plants for treatment of contaminated soil, sludges or ground water, by the removal, degradation, or containment of the pollutants present. Such techniques are generally best suited to sites on which low to moderate levels of contamination are present fairly close to the surface and in a relatively shallow band. Within these general constraints, phytoremediation can be used in the remediation of land contaminated with a variety of substances including certain metals, pesticides, solvents and various organic chemicals.

Metal Phytoremediation

The remediation of sites contaminated with metals typically makes use of the natural abilities of certain plant species to remove or stabilise these chemicals by means of bioaccumulation, phytoextraction, rhizofiltration or phytostabilisation.

Phytoextraction

The process of phytoextraction involves the uptake of metal contaminants from within the soil by the roots and their translocation into the above-ground regions of the plants involved. Certain species, termed hyperaccumulators, have an innate ability to absorb exceptionally large amounts of metals compared to most ordinary plants, typically 50–100 times as much (Chaney *et al.*, 1997; Brooks *et al.*, 1998) and occasionally considerably more. The original wild forms are often found in naturally metal-rich regions of the globe where their unusual ability is an evolutionary selective advantage. Ten years ago, the best candidates for removal by phytoextraction were copper, nickel and zinc, since these are the metals most readily taken up by the majority of the varieties of hyperaccumulator plants and they arguably still remain the metal species most often addressed in this way. New varieties of transgenic plants are also appearing which have been shown to improve performance and may ultimately offer new avenues for real-world, small scale applications in future (Gyulai *et al.*, 2005). Plants which can absorb unusually high amounts of chromium and lead have also been successfully trialled in an attempt to extend the potential applicability of this approach to phytoremediation.

There have also been some breakthroughs in current work, building on earlier attempts to find suitable phytoextractors for nickel and cadmium, with practical field successes in extracting cadmium from paddy fields using the two indica rice cultivars MORETSU and IR-8 (Ibaraki, Kuroyanagi and Murakami, 2009). After two years, the cadmium concentration in the local environment had decreased by 18%, which strongly suggest that phytoextraction using high-Cd accumulating rice varieties could be used very effectively to provide a practical remediation system for moderately cadmium-polluted paddy fields and might be suitable for other similar applications also.

Considerable advances have also been made over the last decade in the phytoextraction of arsenic. Its removal poses a big challenge, since arsenic behaves quite differently from other metal pollutants, typically being found dissolved in the groundwater in the form of arsenite or arsenate, and does not readily precipitate. The application of bipolar electrolysis to oxidise arsenite into arsenate, which reacts with ferric ions from an introduced iron anode, represented a step forward in the wider sphere of remediation, but generally conventional remediation techniques aim to produce insoluble forms of the metal's salts, which, though still problematic, are easier to remove. Clearly, then, the prospect of a specific arsenic-tolerant plant selectively pulling the metal from the soil was long recognised as a highly advantageous potential addition to the bioremediator's tool-kit. The earliest candidate for the role was the Chinese ladder brake fern, *Pteris vittata*, which had then been found to accumulate arsenic in concentrations of 5 g/kg of dry biomass. Growing very rapidly and amassing the metal in its root and stem tissue, it is easy to harvest for contaminant removal, making it eminently suitable from a practical point of view. Subsequent work (Gonzaga, Santos and Ma, 2008) has further established the efficacy of this species for the task, having been shown to be able to accumulate up to 2.3% arsenic in its biomass (Gonzaga, Santos and Ma, 2006). Other plant species have also recently been investigated as potentially useful arsenic phytoextractors, including *Vetiveria zizanioides*, known as Khus grass (Singh *et al.*, 2007) and another fern *Pityrogramma calomelanos* (Gonzaga, Santos and Ma, 2006).

Hyperaccumulation

Hyperaccumulation itself is a curious phenomenon and raises a number of fundamental questions. The previously mentioned pteridophyte, *P. vittata*, tolerates tissue levels of 2.3% arsenic and certain strains of naturally occurring alpine pennycress (*Thlaspi caerulescens*) can bioaccumulate around 1.5% cadmium, on the same dry weight basis. These are wholly exceptional concentrations. Quite how the uptake and the subsequent accumulation are achieved are interesting enough issues in their own right. However, more intriguing is why so much should be taken up in the first place. The hyperaccumulation of copper or zinc, for which there is an underlying certain metabolic requirement can, to some extent, be viewed as the outcome of an over-efficient natural mechanism. The biological basis of the uptake of a completely non-essential metal, however, particularly in such amounts, remains open to speculation at this point. Never-the-less, with a plant like *T. caerulescens* showing a zinc removal rate in excess of 40 kg/ha/year, and suggestions that the entire *Pteris* genus could possibly be used in phytoextraction of arsenic (Gonzaga, Santos and Ma, 2006), their enormous potential value in bioremediation is very clear.

In a practical application, appropriate plants are chosen based on the type of contaminant present, the regional climate and other relevant site conditions. This may involve one or a selection of these hyperaccumulator species, dependent on circumstances. After the plants have been permitted to grow for a suitable length of time, they are harvested and the metal accumulated is permanently removed from the original site of contamination. If required, the process may be repeated with new plants until the required level of remediation has been achieved. One of the criticisms commonly levelled at many forms of environmental biotechnology is that all it does is shift a problem from one place to another. The fate of harvested hyperaccumulators serves to illustrate the point, since the biomass thus collected, which has bioaccumulated significant levels of contaminant metals, needs to be treated or disposed of itself, in some environmentally sensible fashion. Typically the options are either composting or incineration. The former must rely on co-composting additional material to dilute the effect of the metal-laden hyperaccumulator biomass if the final compost is to meet permissible levels; the latter requires the ash produced to be disposed of in a hazardous waste landfill. While this course of action may seem a little un-environmental in its approach, it must be remembered that the void space required by the ash is only around a tenth of that which would have been needed to landfill the untreated soil.

An alternative that has sometimes been suggested is the possibility of recycling metals taken up in this way. There are few reasons, at least in theory, as to why this should not be possible, but much of the practical reality depends on the value of the metal in question. Dried plant biomass could be taken to processing works for recycling and for metals like gold, even a very modest plant content could make this economically viable. By contrast, low value materials, like lead, for example would not be a feasible prospect. At the moment, nickel is probably the best studied and understood in this respect. There has been considerable interest in the potential for biomining the metal out of sites which have been subject to diffuse contamination, or former mines where further traditional methods are no longer practical. The manner proposed for this is essentially phytoextraction and early research seems to support the economic case for drying the harvested biomass and then recovering the nickel. Even where the actual post-mining residue has little immediate worth, the application of phytotechnological measures can still be of benefit as a straightforward clean-up, as recent work in Thailand to improve conditions for rice growing around zinc mines has established (Phaenark *et al.*, 2009).

In the light of pilot scale investigations in Australia, using the ability of eucalyptus trees and certain native grasses to absorb metals from the soil, a similar approach was then tested operationally for the decontamination of disused gold mines (Murphy and Butler, 2002). These sites also often contain significant levels of arsenic and cyanide compounds. Managing the country's mining waste is a major expense, costing in excess of Aus\$30 million/year; developing a successful methodology suitable for deployment on a large scale could prove of great economic advantage to the industry.

The case for metals with intermediate market values is also interesting. Though applying a similar approach to zinc, for instance, might not result in a huge commercial contribution to the smelter, it would be a benefit to the metal production and at the same time, deal rationally with an otherwise unresolved disposal issue.

Clearly, the metallurgists would have to be assured that it was a worthwhile exercise – and with the price of metals on the global market being a notoriously cyclical thing, the dimension of time-sensitivity would also need to be factored into the discussion. The recycling question is, then, a long way from being a workable solution, but potentially it could offer a highly preferable option to the landfill route.

Rhizofiltration

Rhizofiltration is the absorption into, or the adsorption or precipitation onto, plant roots of contaminants present in the soil water. The principle difference between this and the previous approach is that rhizofiltration is typically used to deal with contamination in the groundwater, rather than within the soil itself, though the distinction is not always an easy one to draw. The plants destined to be used in this way are normally brought on hydroponically and gradually acclimatised to the specific character of the water which requires to be treated. Once this process has been completed, they are planted on the site, where they begin taking up the solution of pollutants. Harvesting takes place once the plants have become saturated with contaminants and, as with the phytoextraction, the collected biomass requires some form of final treatment. The system is less widely appreciated than the previous technology, but it does have some very important potential applications. Sunflowers were reported as being successfully used in a test at Chernobyl in the Ukraine, to remove radioactive uranium contamination from water in the wake of the nuclear power station accident.

Phytostabilisation

In many respects, phytostabilisation has close similarities with both phytoextraction and rhizofiltration in that it too makes use of the uptake and accumulation by, adsorption onto, or precipitation around, the roots of plants. On first inspection, the difference between these approaches is difficult to see, since in effect, phytostabilisation does employ both extractive and filtrative techniques. However, what distinguishes this particular phytoremediation strategy is that, unlike the preceding regimes, harvesting the grown plants is not a feature of the process. In this sense, it does not remove the pollutants, but immobilises them, deliberately concentrating and containing them within a living system, where they subsequently remain. The idea behind this is to accumulate soil or ground water contaminants, locking them up within the plant biomass or within the rhizosphere, thus reducing their bioavailability and preventing their migration off site. Metals do not ultimately degrade, so it can be argued that holding them in place in this way is the best practicable environmental option for sites where the contamination is low, or for large areas of pollution, for which large scale remediation by other means would simply not be possible.

A second benefit of this method is that on sites where elevated concentrations of metals in the soil inhibits natural plant growth, the use of species which have a high tolerance to the contaminants present enables a cover of vegetation to be re-established. This can be of particular importance for exposed sites, minimising the effects of wind erosion, wash off or soil leaching which otherwise can significantly hasten the spread of pollutants around and beyond the affected land itself.

Organic Phytoremediation

A wide variety of organic chemicals are commonly encountered as environmental pollutants including many types of pesticides, solvents and lubricants. Probably the most ubiquitous of these across the world, for obvious reasons, are petrol and diesel oil. These hydrocarbons are not especially mobile, tend to adhere closely to the soil particles themselves and are generally localised within 2 m of the surface. Accordingly, since they are effectively in direct contact with the rhizosphere, they are a good example of ideal candidates for phytoremediation. However, recent work has shown that plant phylogeny is an important influence on both the tolerance of a potential effector species to organic pollutants and its ability to take them up (Collins and Willey, 2009). In addition, there seems to be no correlation between the uptake of organic pollutants and heavy metals; further work on this could prove invaluable for phytoremediation scheme design as well as minimising the risk or failure when restoring previously contaminated sites.

The mechanisms of organic phytoremediation are typically phytodegradation, rhizodegradation and phytovolatilisation.

Phytodegradation

Phytodegradation, which is sometimes known by the alternative name of phytotransformation, involves the biological breakdown of contaminants, either internally, having first been taken up by the plants, or externally, using enzymes secreted by them. Hence, the complex organic molecules of the pollutants are subject to biodegradation into simpler substances and incorporated into the plant tissues. In addition, the existence of the extra-cellular enzyme route has allowed this technique to be successfully applied to the remediation of chemicals as varied as chlorinated solvents, explosives and herbicides. Since this process depends on the direct uptake of contaminants from soil water and the accumulation of resultant metabolites within the plant tissues, in an environmental application, it is clearly important that the metabolites which accumulate are either non-toxic, or at least significantly less toxic than the original pollutant.

Rhizodegradation

Rhizodegradation, which is also variously described as phytostimulation or enhanced rhizospheric biodegradation, refers to the biodegradation of contaminants in the soil by edaphic microbes enhanced by the inherent character of the rhizosphere itself. This region generally supports high microbial biomass and consequently a high level of microbiological activity, which tends to increase the speed and efficiency of the biodegradation of organic substances within the rhizosphere compared with other soil regions and microfloral communities. Part of the reason for this is the tendency for plant roots to increase the soil oxygenation in their vicinity and exude metabolites into the rhizosphere. It has been estimated that the release of sugars, amino acids and other exudates from the plant and the net root oxygen contribution can account for up to 20% of plant photosynthetic activity per year (Foth, 1990), of which denitrifying bacteria, *Pseudomonas spp.*, and general heterotrophs are the principal beneficiaries. In addition, mycorrhizae fungi associated with the roots also play a part in metabolising organic contaminants. This is an important aspect, since they have unique enzymatic pathways that enable the biodegradation of organic substances that could not otherwise be transformed solely by bacterial action. In principle, rhizodegradation is intrinsic remediation enhanced by entirely natural means, since enzymes which are active within 1 mm of the root itself transform the organic pollutants, in a way which, clearly, would not occur in the absence of the plant. Never-the-less, this is generally a much slower process than the previously described phytodegradation.

Phytovolatilisation

Phytovolatilisation involves the uptake of the contaminants by plants and their release into the atmosphere, typically in a modified form. This phytoremediation biotechnology generally relies on the transpiration pull of fast growing trees, which accelerates the uptake of the pollutants in ground water solution, which are then released through the leaves. Thus the contaminants are removed from the soil, often being transformed within the plant before being voided to the atmosphere. One attempt which has been explored experimentally uses a genetically modified variety of the Yellow Poplar, *Liriodendron tulipifera* which has been engineered by the introduction of mercuric reductase gene (*mer A*) as discussed in Chapter 9. This confers the ability to tolerate higher mercury concentrations and to convert the metal's ionic form to the elemental and allows the plant to withstand contaminated conditions, remove the pollutant from the soil and volatise it. The advantages of this approach are clear, given that the current best available technologies demand extensive dredging or excavation and are heavily disruptive to the site.

The choice of a poplar species for this application is interesting, since they have been found useful in similar roles elsewhere. Trichloroethylene (TCE), an organic compound used in engineering and other industries for degreasing, is a particularly mobile pollutant, typically forming plumes which move beneath the soil's surface. In a number of studies, poplars have been shown to be able to volatilise around 90% of the TCE they take up. In part this relates to their enormous hydraulic pull, a property which will be discussed again later in this chapter. Acting as large, solar-powered pumps, they draw water out of the soil, taking up contaminants with it, which then pass through the plant and out to the air.

The question remains, however, as to whether there is any danger from this kind of pollutant release into the atmosphere and the essential factor in answering that must take into account the element of dilution. If the trees are pumping out mercury, for instance, then the daily output and its dispersion rate must be such that the atmospheric dilution effect makes the prospect of secondary effects, either to the environment or to human health, impossible. Careful investigation and risk analysis is every bit as important for phytoremediation as it is for other forms of bioremediation.

Using tree species to clean up contamination has begun to receive increasing interest. Phytoremediation in general tends to be limited to sites where the pollutants are located fairly close to the surface, often in conjunction with a relatively high water table. Research in Europe and the US has shown that the deeply penetrating roots of trees allow deeper contamination to be treated. Once again, part of the reason for this is the profound effect these plants can have on the local water relations.

Hydraulic containment

Large plants can act as living pumps, pulling large amounts of water out of the ground which can be a useful property for some environmental applications, since the drawing of water upwards through the soil into the roots and out through the plant decreases the movement of soluble contaminants downwards, deeper into the site and into the groundwater. Trees are particularly useful in this respect because of their enormous transpiration pull and large root mass. Poplars, for example once established, have very deep tap roots and they take up large quantities of water, transpiring between 200 and 1100 litres daily. In situations where grassland would normally support a water table at around 1.5 m, this action can lead to it being up to 10 times lower. The aim of applying this to a contamination scenario is to create a functional water table depression, to which pollutants will tend to be drawn and from which they may additionally be taken up for treatment. This use of the water uptake characteristics of plants to control the migration of contaminants in the soil is termed hydraulic containment, shown schematically in Figure 7.1, and a number of particular applications have been developed.

Buffer strips are intended to prevent the entry of contaminants into watercourses and are typically used along the banks of rivers, when they are sometimes called by the alternative name of 'riparian corridors', or around the perimeter of affected sites to contain migrating chemicals. Various poplar and willow varieties, for example have shown themselves particularly effective in reducing the wash-out of nitrates and phosphates making them useful as pollution control measures to avoid agricultural fertiliser residues contaminating waterways. Part of the potential of this approach is that it also allows for the simultaneous integration of other of the phytoremediating processes described into a natural treatment train, since as previously stated, all plant-based treatments are aspects of the same fundamental processes and thus part of a cohesive whole.

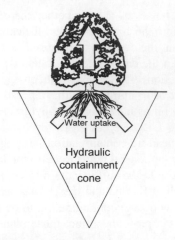

Figure 7.1   *Schematic hydraulic containment*

Another approach sometimes encountered is the production of vegetative caps, which has found favour as a means of finishing off some American landfill sites. The principle involves planting to preventing the downward percolation of rainwater into the landfill and thus minimising leachate production while at the same time reducing erosion from the surface. The method seems to be successful as a living alternative to an impermeable clay or geopolymer barrier. The vegetative cap has also been promoted for its abilities to enhance the biological breakdown of the underlying refuse. In this respect, it may be seen as an applied form of rhizodegradation or even, arguably, of phytodegradation. How effective it is likely to be in this role, however, given the great depths involved in most landfills and the functionally anoxic conditions within them, appears uncertain.

To understand the overall phytoremediation effect of hydraulic containment, it is important to realise that contaminating organics are actually taken up by the plant at lower concentration than they are found *in situ*, in part due to membrane barriers at the root hairs. In order to include this in a predictive mathematical model, the idea of a transpiration stream concentration factor (TSCF) for given contaminants has been developed, defined as TSCF $= 0.75 \exp\{-[(\log K_{ow} - 2.50)2/2.4]\}$, (Burken and Schnoor, 1998) where K_{ow} is the octanol-water partition coefficients. These latter are a measure of the hydrophobia or hydrophilia of a given organic chemical; a log K_{ow} below 1 characterises the fairly soluble, while above 3.5 indicates highly hydrophobic substances.

Thus the uptake rate (U in mg/day) is given by the following equation

$$U = (TSCF)TC$$

Where,

TSCF = transpiration stream concentration factor, as defined
T = transpiration rate of vegetation, l/day
C = concentration in site water, mg/l.

However, it must also be remembered in this context that, should the pollutants not themselves actually be taken up by the plants, then the effect of establishing a hydraulic containment regime will be to increase their soil concentration due to transpiro-evaporative concentration. Thus, the mass of affected water in the contaminant plume reduces, as does the consequent level of dilution it offers and hence, increased localised concentration can result.

The transpiration pull of plants, and particularly tree species, has also sometimes been harnessed to overcome localised water-logging, particularly on land used for agricultural or amenity purposes. To enhance the effect at the point worst affected, the planting regime may involve the establishment of close grouping, which then function as single elevated withdrawal points. The noted ability of poplars to act as solar-powered hydraulic pumps makes them of great potential benefit to this kind of phytotechnological application. Although other plant-based processes could be taking place at the same time to remediate land alongside this to clean up contaminated soils, this particular technique is not itself a type of phytoremediation. Instead, it is an example of the broader bio-engineering possibilities which are offered by the appropriate use of flora species to wider environmental nuisances, which, for some sites, may be the only economic or practicable solution. This may be of particular relevance to heavy soils with poor natural inter-particulate spacing, since laying adequate artificial drainage systems can often be expensive to do in the first place and are frequently prone to collapse once installed.

Another similar example of the use of phytotechnology to overcome nuisance is the bio-bund, which consists of densely planted trees, often willows, on an engineered earthwork embankment. This system has been used successfully to reduce noise pollution from roads, railways and noisy industrial sites, the inter-locking branches acting as a physical barrier to deaden the sound as well as having a secondary role in trapping wind-blown particulates. Depending on the individual site, the bio-bund can be constructed in such a way that it can also act as a buffer strip to control migrating chemical pollution, if required.

Plant Selection

It should be obvious that the major criteria for plant selection are the particular requirements for the method to be employed and the nature of the contaminants involved. For example, in the case of organic phytotransformation this means species of vegetation which are hardy and fast growing, easy to maintain, have a high transpiration pull and transform the pollutants present to non-toxic or less toxic products. In addition, for many such applications, deep rooting plants are particularly valuable.

On some sites, the planting of grass varieties in conjunction with trees, often in between rows of trees to stabilise and protect the soil, may be the best route since they generate a tremendous amount of fine roots near to the surface. This particularly suits them to transforming hydrophobic contaminants such as benzene, toluene, ethylbenzene, xylenes (collectively known as BTEXs) and polycyclic aromatic hydrocarbons (PAHs). They can also be very helpful in controlling wind-blown dust, wash-off and erosion. The selection of appropriate plant species for bio-engineering is not, however, limited solely to their direct ability to treat contaminants, since the enhancement of existing conditions forms as much a part of the potential applications of phytotechnology as bioremediation. For instance, legumes can be of great benefit to naturally nitrogen deficient soils, since they have the ability, via symbiotic root nodule bacteria, to directly fix nitrogen from the atmosphere. With so much to take into consideration in plant selection, the value of a good botanist or agronomist in any interdisciplinary team is clear.

Applications

Phytotechnology has many potentially beneficial land uses, and the applications are still developing, but despite the increasing interest being shown in them, it is probably fair to say that they have not entirely reached their full potential yet. In part this may be due to some of the doubts that have been voiced, suggesting that the beneficial effects of plant utilisation, particularly in respect of phytoremediation, have been over-stated. Some have argued that the reality may range from genuine enhancement to no effect, or even to a negative contribution under certain circumstances and that the deciding factors have more to do with the nature of the site than the plants themselves. In addition, some technologies which have been successfully used on some sites may simply serve to complicate matters on others. One such approach which achieved commercial scale use in the United States, principally for lead remediation, required the addition of chemicals to induce metal take up. Lead normally binds strongly to the soil particles and so its release was achieved by using chelating agents like ethylene diamine tetra acetic acid (EDTA), which were sprayed onto the ground. With the lead rendered biologically available, it can be taken up by plants and hence removed. However, dependent on the character of the site geology, it has been suggested that this could also allow lead to percolate downwards through the soil, and perhaps ultimately into watercourses. While it may well be possible to overcome this potential problem, using accurate mathematical modelling, followed by the establishment of good hydraulic containment as an adjunct to the process, or by running it in a contained biopile, it does illustrate one of the major practical limitations of plant bio-engineering. The potential benefits of phytotechnology for inexpensive, large scale land management are clear, but the relative lack of quantitative field data on its efficacy, especially compared with actively-managed alternative treatment options, is a serious barrier to its wider adoption. In addition, the roles of enzymes, exudates and metabolites need to be more clearly understood and the selection criteria for plant species and systems for various contamination events require better codification. Much research is underway in both public and the private sectors which should ultimately throw considerable light on these issues and allow meaningful comparisons to be drawn once and for all.

One area where phytoremediation could have a particular role to play, and one which might be amenable to early acceptance is as a polishing phase in combination with other clean up technologies. As a finishing process following on from a preceding bioremediation or non-biological method first used to deal with 'hot-spots', plant-based remediation could well represent an optimal low-cost solution. This has already been successful in small scale trials and consequently techniques have been suggested to treat deeply located contaminated ground water by simply pumping to the surface and using it as the irrigant for carefully selected plant species, allowing them to biodegrade the pollutants. The lower levels of site intrusion and engineering required to achieve this would bring clear benefits to both the safety and economic aspects of the remediation operation, if the planned larger scale investigations are found to support the idea.

Aquatic Phyto-Systems (APS)

Aquatic phyto-systems (APS) are principally used to process effluents of one form or another, though manufactured wetlands have been used successfully to remediate some quite surprising soil contaminants, including TNT residues. Though the latter type of application will be discussed in this section, it is probably best considered as an intergrade between the other APS described hereafter and the Terrestrial Phyto-Systems (TPS) of the previous. Many of the aspects of the bio-treatment of sewage and other wastewaters have already been covered in the previous chapter and so will not be restated here. The major difference between conventional approaches to deal with effluents and phytotechnological methods is that the former tend to rely on a faster, more intensively managed and high energy regime, while in general the stabilisation phase of wastewaters in aquatic systems is relatively slow. The influx and exit of effluent into and out of the created wetland must be controlled to ensure an adequate retention period to permit sufficient residence time for pollutant reduction, which is inevitably characterised by a relatively slow flow rate. However, the efficiency of removal is high, typically producing a final treated off-take of a quality which equals, or often exceeds, that of other systems. Suffice it to say that, as is typical of applications of biological processing in general, there are many common systemic considerations and constraints which will obviously affect phyto-systems, in much the same way as they did for technologies which rely on microbial action for their effect.

Many aquatic plant species have the potential to be used in treatment systems and the biological mechanisms by which some of the effects they achieve will already be largely familiar from the preceding discussion of terrestrial systems. There are a number of ways in which APS can be categorised but perhaps the most useful relates to the natural division between algae and macrophytes, which has been adopted, accordingly, here.

Macrophyte Treatment Systems (MaTS)

The discharge of wastewaters into natural watercourses, ponds and wetlands is an ancient and long established practice, though rising urbanisation led to the development of more engineered solutions, initially for domestic sewage and then later, industrial effluents, which in turn for a time lessened the importance of the earlier approach. However, there has been a resurgence of interest in simpler, more natural methods for wastewater treatment and Macrophyte Treatment Systems (MaTS), in particular, have received much attention as a result. While there has, undoubtedly, been a strong upsurge in public understanding of the potential for environmentally harmonious water cleaning *per se*, a large part of the driving force behind the newly found interest in these constructed habitats comes from biodiversity concerns. With widespread awareness of the dwindling number of natural wetlands, often a legacy of deliberate land drainage for development and agricultural purposes, the value of such manufactured replacements has become increasingly apparent. In many ways it is fitting that this should be the case, since for the majority of aquatic macrophyte systems, even those expressly intended as 'monocultures' at the gross scale, it is very largely as a result of their biodiversity that they function as they do.

These treatment systems, shown diagrammatically in Figure 7.2, are characterised by the input of effluent into a reservoir of comparatively much larger volume, either in the form of an artificial pond or an expanse of highly saturated soil held within a containment layer, within which the macrophytes have been established. Less commonly, pre-existing natural features have been used. Although wetlands have an innate ability to accumulate various unwanted chemicals, the concept of deliberately polluting a habitat by using it as a treatment system is one with which few feel comfortable today. A gentle hydraulic flow is established, which encourages the incoming wastewater to travel slowly through the system. The relatively long retention period that results allows adequate time for processes of settlement, contaminant uptake, biodegradation and phytotransformation to take place.

The mechanisms of pollutant removal are essentially the same, irrespective of whether the particular treatment system is a natural wetland, a constructed monoculture or polyculture and independent of whether the macrophytes in question

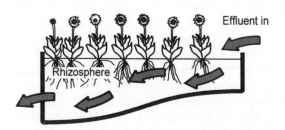

Figure 7.2   *Diagrammatic macrophyte treatment system (MaTS)*

are submerged, floating or emergent species. Both biotic and abiotic methods are involved. The main biological mechanisms are direct uptake and accumulation, performed in much the same manner as terrestrial plants. The remainder of the effect is brought about by chemical and physical reactions, principally at the interfaces of the water and sediment, the sediment and the root or the plant body and the water. In general, it is possible to characterise the primary processes within the MaTS as the uptake and transformation of contaminants by microorganisms and plants and their subsequent biodegradation and biotransformation; the absorption, adsorption and ion exchange on the surfaces of plants and the sediment; the filtration and chemical precipitation of pollutants via sediment contact; the settlement of suspended solids; the chemical transformation of contaminants. It has been suggested that although settlement inevitably causes the accrual of metals, in particular, within the sediment, the plants themselves do not tend to accumulate them within their tissues. While this appears to be borne out, particularly by original studies of natural wetlands used for the discharge of mine washings (Hutchinson, 1975), this does not form any basis on which to disregard the contribution the plants make to water treatment. For one thing, planting densities in engineered systems are typically high and the species involved tend to be included solely for their desired phytoremediation properties, both circumstances seldom repeated in nature. Moreover, much of the biological pollutant abatement potential of the system exists through the synergistic activity of the entire community and, in purely direct terms, this largely means the indigenous microbes. Functionally, there are strong parallels between this and the processes of enhanced rhizospheric biodegradation described for terrestrial applications. While exactly the same mechanisms are available within the root zone in an aquatic setting, in addition, and particularly in the case of submerged vegetation, the surface of the plants themselves becomes a large extra substrate for the attached growth of closely associated bacteria and other microbial species. The combined rhizo- and circum-phyllo- spheres support a large total microbial biomass, with a distinctly different compositional character, which exhibits a high level of bio-activity, relative to other microbial communities. As with rhizodegradation on dry land, part of the reason is the increased localised oxygenation in their vicinity and the corresponding presence of significant quantities of plant metabolic exudates, which, as was mentioned in the relevant earlier section, represents a major proportion of the yearly photosynthetic output. In this way, the main role of the macrophytes themselves clearly is more of an indirect one, bringing about local environmental enhancement and optimisation for remediative microbes, rather than being directly implicated in activities of primary biodegradation. In addition, physico-chemical mechanisms are also at work. The iron plaques which form on the plant roots trap certain metals, notably arsenic (Otte, Kearns and Doyle, 1995), while direct adsorption and chemical/biochemical reactions play a role in the removal of metals from the wastewater and their subsequent retention in sediments.

The ability of emergent macrophytes to transfer oxygen to their submerged portions is a well appreciated phenomenon, which in nature enables them to cope with effective waterlogging and functional anoxia. As much as 60% of the oxygen transported to these parts of the plant can pass out into the rhizosphere, creating aerobic conditions for the thriving microbial community associated with the root zone, the leaf surfaces and the surrounding substrate. This accounts for a significant increase in the dissolved oxygen levels within the water generally and, most particularly, immediately adjacent to the macrophytes themselves.

The aerobic breakdown of carbon sources is facilitated by this oxygen transfer, for obvious reasons, and consequently it can be seen to have a major bearing on the rate of organic carbon biodegradation within the treatment system, since its adequate removal requires a minimum oxygen flux of one and a half times the input BOD loading. Importantly, this also makes possible the direct oxidation of hydrogen sulphide (H_2S) within the root zone and, in some cases, iron and manganese.

While from the earlier investigations mentioned on plant/metal interactions (Hutchinson, 1975) their direct contribution to metal removal is small, fast growing macrophytes have a high potential uptake rate of some commonly encountered effluent components. Some kinds of water hyacinth, *Eichhornia spp.*, for example can increase their biomass by $10 \, g/m^2$/day under optimum conditions, which represents an enormous demand for nitrogen and carbon from their environment. The direct uptake of nitrogen from water by these floating plants gives them an effective removal potential which approaches 6000 kg/ha/year and this, coupled with their effectiveness in degrading phenols and in reducing copper, lead, mercury, nickel and zinc levels in effluents, explains their use in bio-engineered treatment systems in warm climates.

Emergent macrophytes are also particularly efficient at removing and storing nitrogen in their roots, and some can do the same for phosphorus. However, the position of this latter contaminant in respect of phyto-treatment in general is less straightforward. In a number of constructed wetland systems, though the overall efficacy in the reduction of BOD, and the removal of nitrogenous compounds and suspended solids has been high, the allied phosphorus components have been dealt with much less effectively. This may be of particular concern if phosphorus-rich effluents are to be routinely treated and there is a consequent risk of eutrophication resulting. It has been suggested that, while the reasons for this poor performance are not entirely understood, nor is it a universal finding for all applications of phyto-treatment, it may be linked to low root zone oxygenation in slow-moving waters (Heathcote, 2000). If this is indeed the case, then the preceding discussion on the oxygen pump effect of many emergent macrophytes has clear implications for bio-system design.

As has been established earlier, associated bacteria play a major part in aquatic plant treatment systems and microbial nitrification and denitrification processes are the major nitrogen-affecting mechanisms, with anaerobic denitrification, which typically takes place in the sediment, causing loss to atmosphere, while aerobic nitrification promotes and facilitates nitrogenous incorporation within the vegetation. For the effective final removal of assimilated effluent components, accessibly harvestable material is essential, and above-water, standing biomass is ideal.

The link between the general desire for biodiversity conservation and the acceptability of created wetlands was mentioned earlier. One of the most important advantages of these systems is their potential to create habitats not just for 'popular' species, like waterfowl, but also for many less well known organisms, which can be instrumental in bolstering the ecological integrity of the area. This may be of particular relevance in industrial or urban districts. At the same time, they can be ascetically pleasing, enhancing the landscape while performing their function. These systems can have relatively low capital costs, but inevitably every one must be heavily site specific, which means many aspects of the establishment financing are variable. However, the running costs are generally significantly lower than for comparable conventional treatment operations of similar capacity and efficacy. In part the reason for this is that once properly set up, a well designed and constructed facility is almost entirely self-maintaining. However, the major contribution to low operational overheads comes from the system's low energy requirements, since gravity drives the water flow and all the remediating organisms are ultimately solar powered, either directly or indirectly, via the photosynthetic action of the resident autotrophe community.

Aside from cost and amenity grounds, one major positive feature is that the effluent treatment itself is as good or better than that from conventional systems. When correctly designed, constructed, maintained and managed, plant-based treatment is a very efficient method of ameliorating wastewaters from a wide range of sources and in addition, is very tolerant of variance in organic loadings and effluent quality, which can cause problems for some of the alternative options. In addition, phyto-systems can often be very effective at odour reduction, which is often a major concern for the producers and processors of effluents rich in biodegradable substances.

Invariably, the better designed, the easier the treatment facility is to manage and in most cases, 'better' means simpler in practice, since this helps to keep the maintenance requirement to a minimum and makes maximum use of the existing topography and resources. Provision should also be made for climatic factors and most especially, for the possibility of flooding or drought. It is imperative that adequate consideration is given to the total water budget at the project planning stage. Although an obvious point, it is important to bear in mind that one of the major constraints on the use of aquatic systems is an adequate supply of water throughout the year. While ensuring this is seldom a problem for temperate lands, for some regions of the world it is a significant concern. Water budgeting is an attempt to model the total requirement, accounting for the net overall in- and outputs, together with the average steady-state volume resident within the system in operation. Thus, effluent inflow, supplementary 'clean' water and rainfall need to be balanced against off-take, evaporative and transpirational losses and the demands of the intended retention time required to treat the particular contaminant profile of a given wastewater. One apparent consideration in this process is the capacity of the facility. Determining the 'required' size for a treatment wet-land is often complicated by uncertainty regarding the full range of wastewater volumes and component character likely to be encountered over the lifetime of the operation. The traditional response to this is to err on the side of caution and oversize, which, of course, has inevitable cost implications, but in addition, also affects the overall water budget. If the effluent character is known, or a sample can be obtained, its BOD can be found and it is then a relatively simple procedure to use this to calculate the necessary system size. However, this should only ever be taken as indicative. For one thing, bio-engineered treatment systems typically have a lifespan of 15–20 years and the character of the effluent being treated may well change radically over this time, particularly in response to shifts in local industrial practice or profile. In addition, though BOD assessment is a useful point of reference, it is not a uniform indicator of the treatment require-ments of all effluent components. For the bioamelioration process to proceed efficiently, a fairly constant water level is necessary. Although the importance of this in a drought scenario is self-evident, an unwanted influx of water can be equally damaging, disturbing the established equilibrium of the wetland and pushing contaminants through the system before they can be adequately treated. Provision to include sufficient supplementary supplies, and exclude surface water, is an essential part of the design process.

One aspect of system design which is not widely appreciated is the importance of providing a substrate with the right characteristics. A number of different materials have been used with varying degrees of success, including river sands, gravels, pulverised clinker, soils and even waste-derived composts, the final choice often being driven by issues of local availability. The main factors in determining the suitability of any given medium are its hydraulic permeability and absorbance potential for nutrients and pollutants. In the final analysis, the sub-strate must be able to provide an optimum growth medium for root development while also allowing for the uniform infiltration and through-flow of wastewater. A hydraulic permeability of between 10^{-3} and 10^{-4} m/s is generally accepted as ideal, since lower infiltration tends to lead to channelling and flow reduction, both of which severely restrict the efficiency of treatment. In addition, the chemical nature of the chosen material may have an immediate bearing on system efficacy. Soils with low inherent mineral content tend to encourage direct nutrient uptake to make good the deficiency, while highly humeric soils have been shown to have the opposite effect in some studies. The difficulties sometimes encountered in relation to phosphorus removal within wetland systems have been mentioned earlier. The character of the substrate medium can have an important influence on the uptake of this mineral, since the physico-chemical mechanisms responsi-ble for its abstraction from wastewater in an aquatic treatment system relies on the presence of aluminium or iron within the rhizosphere. Obviously, soils with high relative content of these key metals will be more effective at removing the phosphate component from effluent, while clay-rich substrates tend to be better suited to lowering heavy metal content.

Engineered reed beds are probably the most familiar of all MaTS, with several high profile installations in various parts of the globe having made the technology very widely accessible and well appreciated. This approach has been successfully applied to a wide variety of industrial effluents, in many different climatic conditions and has currently been enjoying considerable interest as a 'green' alternative to septic tanks for houses not joined up to mains sewerage. At its heart is the ability of reeds, often established as monocultures of individual species, or sometimes as oligocultures of a few, closely related forms, to force oxygen down into the rhizosphere, as has been previously discussed. Many examples feature *Phragmites* or *Typha* species, which appear to be particularly good exponents of the oxygen pump, while simultaneously able to support a healthy rhizospheric microfloral complement and provide a stable root zone lattice for associated bacterial growth and physico-chemical processing of rhizo-contiguous contaminants. Isolated from the surrounding ground by an impermeable clay or polymer layer, the reed bed is almost the archetypal emergent MaTS.

The mechanisms of action are shown in Figure 7.3 and may be categorised as surface entrapment of any solids or relatively large particulates on the growing medium or upper root surface. The hydraulic flow draws the effluent down through the rhizosphere, where the biodegradable components come into direct contact with the root zone's indigenous micro-organisms, which are stimulated to enhanced metabolic activity by the elevated aeration and greater nutrient availability. There is a net movement of oxygen down through the plant and a corresponding take up by the reeds of nitrates and minerals made accessible by the action of nitrifying and other bacteria.

These systems are very efficient at contamination removal, typically achieving 95% or better remediation of a wide variety of pollutant substances, as demonstrated in Table 7.1, which shows illustrative data on the amelioration of landfill leachates by this system.

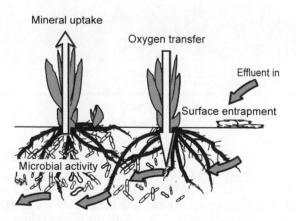

Figure 7.3   *Diagrammatic root zone activity*

Table 7.1   Typical analysis for leachate treatment by reed-bed systems

Component	Inlet conc	Outlet conc	% reduction
Chlorobenzene	83.0	<1.0	>98.7
1,2-Dichlorobenzene	2.0	<1.0	>50
1,3-Dichlorobenzene	3.6	<1.0	>72
1,4-Dichlorobenzene	9.8	<1.0	>89
1,2,4-Trichlorobenzene	0.22	<0.05	>77
α-Hydrocarbons	0.11	0.037	>66
β-Hydrocarbons	0.22	<0.005	>97.7
γ-Hydrocarbons	0.45	0.026	>94
δ-Hydrocarbons	1.0	0.21	>79
ε-Hydrocarbons	1.3	0.31	>76
2-Chlorophenol	3.7	0.84	>77
3,4-Chlorophenol	0.63	<0.5	>92
2,5-Dichlorophenol	5.0	0.56	>88
3,4-Dichlorophenol	4.9	<0.5	>98.9
3,5-Dichlorophenol	0.93	<0.5	>94
2,3,4,6-Tetrachlorophenol	6.4	<1.0	>84
Pentachlorophenol	2.7	<1.0	>62
2,2',5,5'-Penta-PCB	0.11	<0.05	>54
2,2',4,4',5,5'-Hexa-PCB	0.046	<0.025	>45
Naphthalene	17.0	<0.05	>99.7
1-Methylnaphthalene	1.7	<0.05	>97
2-Methylnaphthalene	0.58	<0.05	>91
Acenaphthalene	2.2	<0.05	>97
Fluorene	3.0	<0.05	>98
Phenanthrene	2.7	<0.05	>98
Anthracene	0.4	<0.05	>87
Pyrene	0.29	<0.05	>82
Fluoranthene	0.5	<0.05	>90
Trichloroethene	1.8	<0.05	>97
Tetrachloroethene	3.1	0.11	>96
Benzene	52.0	<1.0	>98
Toluene	1.8	<1.0	>44
Ethylbenzene	16.0	<1.0	>93
m-/p-Xylene	4.1	<1.0	>75
o-Xylene	3.1	<1.0	>67
1,2,4-Trimethylbenzene	3.1	<1.0	>67

Concentrations are expressed in micrograms per litre (after Kickuth).

Never-the-less, reed beds and root zone treatment techniques in general are not immune from a range of characteristic potential operational problems, which can act to limit the efficacy of the process. Thus, excessive water-logging, surface run-off, poor or irregular substrate penetration and the development of preferential drainage channels across the beds may all contribute to a lessening of the system's performance, in varying degrees.

Nutrient Film Techniques (NFTs)

An alternative approach to the use of aquatic macrophytes, which was tried experimentally, involved growing plants on an impermeable containment layer, in a thin film of water. In this system, the wastewater flowed directly over the root mass, thereby avoiding some of the mass transfer problems sometimes encountered by other aquatic phyto-treatment regimes. Though the early work indicated that it had considerable potential for use in the biological treatment of sewage and other nutrient-rich effluents, it does not appear to have been developed further and little is known as to the conditions which govern its successful practical application. One interesting aspect which did, however, emerge at the time was that the cultivation system could also be extended to most terrestrial plants. Subsequently, the potential for developing Nutrient Film Techniques (NFTs) into a relatively simple and inexpensive means to treat municipal wastewaters sufficiently to meet discharge standards has indeed been shown for a number of such species, including *Datura innoxia* (Vaillant *et al.*, 2003) *Chrysanthemum cinerariaefolium* (Vaillant *et al.*, 2002), a variety of commercial roses (Monnet *et al.*, 2002) and both *Digitalis lanata* and *purpurea* (Vaillant *et al.*, 2004). This may ultimately pave the way for the further development of such land-based phyto-treatments in the future.

Algal Treatment Systems (ATS)

Algae have principally been employed to remove nitrogen and phosphorus from wastewaters, though some organic chemicals can also be treated and there has been renewed interest in making use of their efficient carbon sequestration potential.

Effluent treatment

Algal effluent treatment systems work on the basis of functional eutrophication and rely on a dynamic equilibrium between the autotrophic algae themselves and the resident heterotrophic bacteria, which establishes a two stage biodegradation/assimilation process, as shown in Figure 7.4. In effect this is an ecological microcosm in which organic contaminants present in the wastewater are biologically decomposed by the aerobic bacteria, which make use of oxygen provided by algal photosynthesis, while the algae grow using the nutrients produced by this bacterial breakdown, and photosynthesise producing more oxygen.

Though the process is self-sustaining, it is also self-limiting and left to proceed unchecked, will result in the well-appreciated characteristic eutrophic stages leading to the eventual death of all component organisms, since true climactic balance is never achieved in the presence of continuously high additional nutrient inputs. The removal of excess algal and bacterial biomass is, therefore, an essential feature, vital to maintaining the system's efficiency.

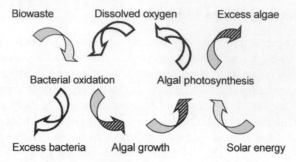

Biowaste   Dissolved oxygen   Excess algae

Bacterial oxidation   Algal photosynthesis

Excess bacteria   Algal growth   Solar energy

Figure 7.4   *Algal and bacterial equilibrium*

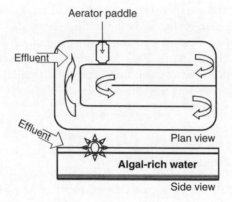

Aerator paddle

Effluent

Effluent   Plan view

Algal-rich water

Side view

Figure 7.5   *High rate algal pond*

Of all the engineered algal systems for effluent, the high rate algal pond (HRAP) is one of the most efficient and represents a good illustration of this use of phytotechnology. Figure 7.5 shows a typical example.

The system consists of a bioreactor cell in the form of a relatively shallow reservoir, typically between 0.2 and 0.6 m deep, with a length to width ratio of 2 : 1 or more, the idea being to produce a large surface area to volume ratio. The void is divided with internal baffles forming walls, to create a channel through which the effluent flows. A mechanically driven paddle at the end nearest to the effluent input both aerates and drives the wastewater around the system. These ponds are not sensitive to fluctuations in daily feed, either in terms of quantity or quality of effluent, providing that it is fundamentally of a kind suitable for this type of treatment. Consequently, they may be fed on a continuous or intermittent basis. The main influences which affect the system's performance are the composition of the effluent, the efficiency of mixing, the retention time, the availability and intensity of light, pond depth and temperature. The latter two factors are particularly interesting since they form logical constraints on the two groups of organisms responsible for the system's function, by affecting the autotrophe's ability to photosynthesise and the heterotrophe's to respire. While a deeper cell permits greater resident biomass, thus elevating the numbers of micro-organisms available to work on the effluent, beyond a certain limit, the law of diminishing returns applies in respect of light available to algae in the lower reaches. Warmer temperatures increase metabolic activity, at least within reason, and the rate of straightforward chemical reactions doubles per $10\,^{\circ}C$ rise, but at the same time, elevated water temperatures have a reduced oxygen carrying capacity which affects the bacterial side of the equilibrium mentioned earlier. As with so much of environmental biotechnology, a delicate balancing act is required.

After a suitable retention period, which again depends on the character of the effluent, the design and efficacy of the treatment pond and the level of clean-up required, the water is discharged for use or returned to watercourses. Obviously, after a number of cycles, algal and bacterial growth in a functionally eutrophic environment would, as discussed earlier in the section, begin to inhibit, and then eventually arrest, the bio-treatment process. By harvesting the algal biomass, not only are the contaminants, which to this point have been merely biologically isolated, physically removed from the system, but also a local population depression is created, triggering renewed growth and thus optimised pollutant uptake. The biomass recovered in this way has a variety of possible uses, of which composting for ultimate nutrient reclamation is without doubt the most popular, though various attempts have also been made to turn the algal crop into a number of different products, including animal feed and insulating material.

Carbon sequestration

Their use as a carbon sink is a simpler process, only requiring the algae themselves. However, even as a functional algal monoculture, just as with the joint algal/bacterial bio-processing for effluents, without external intervention to limit the standing burden of biomass within the bioreactor, reduced efficiency and, ultimately, system collapse is inevitable.

In nature, huge amounts of many elements are held in global reservoirs, regulated by biogeochemical cycles, driven by various interrelated biological and chemical systems. For carbon, a considerable mass is held in organic and inorganic oceanic stores, with the seas themselves being dynamic and important component parts of the planetary carbon cycle. Marine phytoplankton utilise carbon dissolved in the water during photosynthesis, incorporating it into biomass and simultaneously increasing the inflow gradient from the atmosphere. When these organisms die, they sink, locking up this transient carbon and taking it out of the upper oceanic 'fast' cycle into the 'slow' cycle, which is bounded by long-term activities within the deep ocean sediments. In this respect, the system may be likened to a biological sequestration pump, effectively removing atmospheric CO_2 from circulation within the biosphere on an extended basis. The number, mass and extent of phytoplankton throughout the world's seas thus provide a carbon-buffering capacity on a truly enormous scale, the full size of which has only really become apparent within the last 20 years, with the benefit of satellite observation and most particularly the recent use of the Moderate Resolution Imaging Spectroradiometer (MODIS) aboard NASA's *Aqua* vehicle (Kwiatkowska and McClain, 2009).

In the century since its effectiveness as a means of trapping heat in the atmosphere was first demonstrated by the Swedish scientist, Svante Arrhenius, the importance of reducing the global carbon dioxide emissions has come to be widely appreciated. The increasing quantities of coal, oil and gas that are burnt for energy has led to CO_2 emissions worldwide becoming more than 10 times higher than they were in 1900 and there is around 40% more CO_2 in the air, currently 387 ppm according to the US National Oceanic and Atmospheric Administration, than before the Industrial Revolution. Carbon dioxide has been implicated in over 80% of global warming and according to atmospheric monitoring and analysis of samples of the Antarctic ice, the world today has higher levels of greenhouse gases than at any time in the past 650 000 years. The UN Intergovernmental Panel on Climate Change has warned that immediate action is required to prevent further atmospheric increases above today's level. In the absence of swift and effective measures to control the situation, by 2100 they predict that carbon dioxide concentrations will rise to 550 ppm on the basis of the lowest emission model, or over 830 ppm in the highest. Even in the aftermath of the 2009 UN Copenhagen summit and the so-called 'climate-gate' controversy, such figures are hard to ignore.

In 1990, over 95% of the Western industrialised nations' emissions resulted from burning fossil fuels for energy, with the 25% of the world's population who live in these countries consuming nearly 80% of the energy produced globally. Unsurprisingly, energy industries account for the greatest share (36%) of carbon dioxide emissions, a large 1000 MW coal-fired power station releases something in the region of $5^1/_2$ million tonnes of CO_2 annually. Clearly, the current focus on reducing fossil fuel usage, and on minimising the emissions of carbon dioxide to the atmosphere, is important. In one sense, the most straightforward solution to the problem is simply to stop using fossil fuels altogether. However, this is a rather simplistic view and just too impractical. While great advances have been made in the field of renewable energy, a wholesale substitution for gas, coal and oil is not possible at this time if energy usage is to continue at an unabated rate. The potential role of existing non-fossil fuel technology to bridge the gap between the current status quo and a future time, when renewables meet the needs of mankind, is a vital one. However, it remains ridiculous to pretend that this can be achieved overnight, unless the 'global village' really is to consist of just so many mud huts.

In many respects, here is another case where, if we cannot do the most good, then perhaps we must settle for doing the least harm and the application of phytotechnology stands as one very promising means by which to achieve this goal. The natural contribution of algal photosynthesis to carbon sequestration has already been alluded to and the use of these organisms in an engineered system to reduce CO_2 releases, simply capitalises on this same inherent potential in an unaltered way.

There have been attempts to commercialise the benefits of algae as carbon sinks. In the early 1990s, two prototype systems were developed in the UK, aimed at the reduction of CO_2 emissions from various forms of existing combustion processes. The BioCoil was a particularly interesting integrated approach, removing carbon dioxide from generator emissions and deriving an alternative fuel source in the process. The process centred on the use of unicellular algal species in a narrow, water containing, spiral tube made of translucent polymer, through which the exhaust gases from the generator was passed. The carbon dioxide rich waters provided the resident algal with optimised conditions for photosynthesis which were further enhanced by the use of additional artificial light. The algal biomass recovered from the BioCoil reactor was dried, and being unicellular, the effective individual particle size tended to the dimensions of diesel injection droplets, which, coupled with an energy value roughly equivalent to medium grade bituminous coal at 25 MJ/kg, makes it ideal for use in a suitable engine without further modification. Despite early interest, the system does not appear to have been commercially adopted or developed further.

Around the same time, another method was also suggested by one of the authors. In this case, it was his intent specifically to deal with the carbon dioxide produced when biogas, made either at landfill sites or anaerobic digestion plants, was flared or used for electricity generation. Termed the algal cultivation system and carbon sink (ACSACS), it used filamentous algae, growing as attached biofilter elements on a polymeric lattice support. CO_2 rich exhaust gas was passed into the bottom of a bioreactor vessel, containing the plastic filter elements in water, and allowed to bubble up to the surface through the algal strands as shown in the following Figure 7.6.

Again, this approach to carbon sequestration was based on enhanced intra-reactor photosynthesis, the excess algal biomass being harvested to ensure the

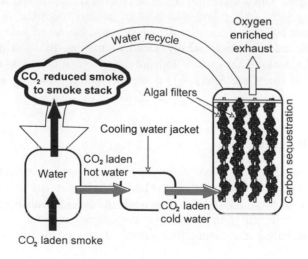

Figure 7.6 *Schematic ACSACS*

ongoing viability of the system, with the intention of linking it into a composting operation to achieve the long-term carbon lock-up desired. The ACSACS
though performing well at both bench and small pilot scale, never attained industrial adoption though remaining an interesting possible adjunct to the increasing
demand for methane flaring or utilisation at landfills.

A similar idea emerged again about 10 years later, with a system being developed by Ohio University, which, in a perfect example of selecting an organism
from an extreme environment to match the demands of a particular man-made
situation, utilised thermophilic algae from hot springs in Yellowstone National
Park. In this process, which received a $1 million grant from the US Department
of Energy, smoke from power stations was to be diverted through water to permit
some of the CO_2 to be absorbed and the hot, carbonated water produced then
flowed through an algal filter formed on vertical nylon screens.

This design, which was essentially similar to the earlier ACSACS, enables
the largest possible algal population per unit volume to be packed into the filter
unit, though like the previously described HRAP, light is a limiting factor, since
direct sunlight will only penetrate through a few feet of such an arrangement.
However, it was claimed that these carbon bio-filters could remove up to 20% of
the carbon dioxide, which would, of course, otherwise be released to atmosphere.
This made solving the problem something of a priority. One solution which was
put forward involved the use of a centralised light collector, connected to a series
of fibre-optic cables linked to diffusers within the vessels to provide adequate
illumination within the filters. An alternative approach was also suggested to use
large artificial lakes, but this would have required a much larger land bank to
produce the same effect, since they would have had to have been very shallow
by comparison. It was also suggested that cooling the carbonated water first, a
feature of both the BioCoil and ACSACS, would have allowed normal mesophilic
algae to be used, which take up CO_2 more efficiently. In the end this technique
seems to have been little more successful in gaining industrial or commercial
acceptance than either of the earlier British systems. However, with both carbon-
consciousness and fears over energy security growing, the vast potential that
algae have for locking up carbon and providing a non-contentious alternative
form of biofuel (see Case Study 7.1) it seems unlikely that new ideas will not
continue to emerge.

Pollution Detection

One final application of phytotechnology to the environmental context to
consider involves the possible use of plants in a variety of industrial sectors
as pollution detectors. The aim is principally to provide valuable information
about the toxicological components of contamination from a wide variety
of sources, including the automotive, chemical and textile industries. Unlike

biosensors, which tend to be designed around isolated biochemical reactions, in this approach, the plants are used as entire biological test systems. Moreover, unlike conventional chemical analytical methods which produce quantifiable, numeric measurements, the varieties used have been selected for their abilities to identify contaminants by reacting to the specific effects these substances have on the plant's vital functions. Thus, by directing the focus firmly onto the obvious and discernable biological consequences of the pollutants and then codifying this into a diagnostic tool, the assessment process is made more readily available to a wider range of those who have an interest in pollution control.

The development of this technology is still in its relatively early stages, but it would appear to open up the way for a controllable method to determine pollutant effects. It seems likely that they will be of particular value as early detection systems in the field, since they are functional within a broad range of pH and under varied climatic conditions. An additional benefit is that they are responsive to both long-term pollution or incidental spillages and can be applied to either laboratory or on-site investigations to monitor air, soil or water, even on turbid or coloured samples, which often cause anomalous readings with spectrophotometric test methods. Moreover, since they are based around gross observable effects, they may offer significant opportunity for effective pollution monitoring for remote areas and parts of the world where skilled laboratory practitioners are in short supply.

Closing Remarks

The scope of opportunities available to phytotechnological interventions are, then, wide ranging, and much of their potential still remains to be explored. It seems likely that the increasingly prioritised initiatives to find low-cost systems to bring about effective remediation, effluent control and carbon sequestration will continue to favour their uptake in the coming years. However, it is unavoidably true that the adoption of any plant-utilising bioengineering applications will depend as much on local modalities as on the actual state of the biotechnologies themselves. In this respect, phytotechnology has much in its favour. For one thing, it has the enormous benefit of virtually assured universal public acceptance, which is rare for any biotechnology. Commercially, it is a relatively low intervention, highly 'green' and thoroughly non-contentious approach to environmental management, which has a strong potentially positive contribution to corporate image, with a relatively low negative influence on the balance sheet. The great advantage that almost all plant-based systems bring to biological engineering is the tremendous energy saving represented by their solar-powered nature. This, combined with their essentially integrated and intrinsically complex array of metabolic mechanisms makes a variety of plant species extremely useful in an environmental context, and typically without any additional need for modification. Given their inherent flexibility, acceptability, efficiency and cost-effectiveness, it is difficult to imagine that phytotechnological systems will not be further developed and continue to be more widely adopted in the future.

References

Brooks, R., Chambers, M., Nicks, L. and Robinson, B. (1998) Phytomining. *Trends in Plant Science*, **3**, 359–362.

Burken, J. and Schnoor, J. (1998) Predictive relationships for uptake of organic contaminants by hybrid poplar trees. *Environmental Science and Technology*, **32**, 3379–3385.

Chaney, R., Malik, M., Li, Y. *et al.* (1997) Phytoremediation of soil metals. *Current Opinion in Biotechnology*, **8**, 279–284.

Collins, C.D. and Willey, N.J. (2009) Phylogenetic variation in the tolerance and uptake of organic contaminants. *International Journal of Phytoremediation*, **11** (7), 623–639.

Foth, H. (1990) *Fundamentals of Soil Science*, 8th edn, John Wiley & Sons, Inc., New York.

Gonzaga, M.I.S., Santos, J.A.G. and Ma, L.Q. (2006) Arsenic phytoextraction and hyperaccumulation by fern species. *Scientia Agricola*, **63** (1).

Gonzaga, M.I.S., Santos, J.A.G. and Ma, L.Q. (2008) Phytoextraction by arsenic hyperaccumulator *Pteris vittata* L. from six arsenic-contaminated soils: repeated harvests and arsenic redistribution. *Environmental Pollution*, **154** (2), 212–218.

Gyulai, G., Humphreys, M.O., Bittsanszky, A. *et al.* (2005) AFLP analysis and improved phytoextraction capacity of transgenic gshI-poplar clones (*Populus* × *canescens*) for copper in vitro. *Zeitschrift fur Naturforschung C*, **60** (3-4), 300–306.

Heathcote, I. (2000) *Artificial Wetlands for Wastewater Treatment*, Prentice Hall and Pearson.

Hutchinson, G. (1975) *A Treatise of Limnology*, Limnological Botany, Vol. III, John Wiley & Sons, Inc., New York.

Ibaraki, T., Kuroyanagi, N. and Murakami, M. (2009) Practical phytoextraction in cadmium-polluted paddy fields using a high cadmium accumulating rice plant cultured by early drainage of irrigation water. *Soil Science and Plant Nutrition*, **55** (3), 421–427.

Kwiatkowska, E.J. and McClain, C.R. (2009) Evaluation of SeaWiFS, MODIS Terra and MODIS Aqua coverage for studies of phytoplankton diurnal variability. *International Journal of Remote Sensing*, **30** (24), 6441–6459.

Monnet, F., Vaillant, N., Hitmi, A. *et al.* (2002) Treatment of domestic wastewater using the nutrient film technique (NFT) to produce horticultural roses. *Water Research*, **36** (14), 3489–3496.

Murphy, M. and Butler, R. (2002) Eucalyptus could be worth weight in gold. *Chemistry and Industry*, **1**, 7.

Otte, M., Kearns, C. and Doyle, M. (1995) Accumulation of arsenic and zinc in the rhizosphere of wetland plants. *Bulletin of Environmental Contamination Toxicology*, **55**, 154–161.

Phaenark, C., Pokethitiyook, P., Kruatrachue, M. and Ngernsansaruay, C. (2009) Cd and Zn accumulation in plants from the Padaeng Zinc Mine area. *International Journal of Phytoremediation*, **11** (5), 479–495.

Singh, S.K., Juwarkar, A., Kumar, S. *et al.* (2007) Effect of amendment on phytoextraction of arsenic by *Vetiveria Zizanioides* from soil. *International Journal of Environmental Science and Technology*, **4** (3), 339–344.

Vaillant, N., Monnet, F., Sallanon, H. *et al.* (2003) Treatment of domestic wastewater by an hydroponic NFT system. *Chemosphere*, **50** (1), 121–129.

Vaillant, N., Monnet, F., Sallanon, H. *et al.* (2004) Use of commercial plant species in a hydroponic system to treat domestic wastewaters. *Journal of Environmental Quality*, **33**, 695–702.

Vaillant, N., Monnet, F., Vernay, P. *et al.* (2002) Urban wastewater treatment by a nutrient film technique system with a valuable commercial plant species (*Chrysanthemum cinerariaefolium*). *Environmental Science and Technology*, **36** (9), 2101–2106.

Case Study 7.1   Algal Biofuel Production (USA)

Algae have an obvious attraction when it comes to biofuel production. While terrestrial plants such as soy, canola and palm yield around 470, 1500 and 5700 l of oil per hectare respectively, some kinds of algae can provide approaching 20 000. The problem in realising this potential tends to come down to the demands for resources and land necessary to make it possible – and in the case of enclosed bioreactors, the cost of construction and operation too.

As part of its 'Spaceship Earth' project, NASA has come up with the idea of huge plastic osmotic containers, filled with sewage and floating at sea, in which to grow algae. The concept is a simple but effective one. NASA has been developing semi-permeable 'forward-osmosis membranes' for some time to recycle water on future long-haul space missions. By using them to create their offshore membrane enclosures for growing algae (OMEGA), additional internal fresh water can safely escape, without the danger of ingress into the system of the external salt water. In addition, the nature of the material also allows for the necessary gaseous exchange, while temperature control – a major cost element in land-based enclosed bioreactors – is provided by the natural buffer effect of the sea's specific heat capacity. Wave action will help keep the system mixed, avoiding the need for paddles, motors and an external power supply.

In addition to the eventual potential fuel production from the harvested algae, the OMEGA system also promises sustainable wastewater treatment, fertiliser production and carbon sequestration. Significantly, given the environmental question mark hanging over some forms of terrestrial bio-energy, it entirely circumvents any necessity to compete with food production for increasingly scarce land and water resources.

Case Study 7.2   Sustainable Engineered Phytoremediation (Dhulikhel, Nepal)

Many practical applications of phytotechnology owe their success to the effective simplicity of their design, a factor which makes them particularly useful in areas short on resources, skilled labour or traditional infrastructure. As a result, engineered phytoremediation in particular can enable reproducibly sustainable development projects to be undertaken in regions where it would be difficult to provide more conventional solutions.

After a series of public health incidents which were shown to be linked to the existing largely *ad hoc* arrangements for sewage, the Dhulikhel municipality – the smallest in Nepal – constructed a community-based wastewater treatment plant, utilising constructed wetlands to return cleaned water to the Punyamata River. Built at Srikhandapur, in the Kavre district, and said to be the first scheme of its kind in Southern Asia, six 175 m^3 horizontal reed beds lie at its core, with two 75 m^3 biogas reactors providing cooking gas for the population. The sludge solids produced are returned to beneficial use as agricultural fertiliser to support the local economy.

The project was instigated as a co-operative venture between the United Nations Human Settlements Programme (UN-HABITAT), the Environment and Public Health Organization (ENPHO), the Dhulikhel municipality and the people themselves. Crucially, in addition to providing direct technical assistance, ENPHO also undertook local training, to ensure that the reed beds will be able to continue to function effectively in future. The success of this approach has led Nepal's Ministry of Physical Planning and Works to consider feasibility studies of the biogas/constructed wetland model for other communities and cities.

It is a particularly good example of the simplicity, sustainability and potential for easy replication that phytotechnology can offer. However, perhaps more importantly, it makes the point very clearly that 'biotech' solutions do not automatically have to be technologically complex, excessively engineered or prohibitively expensive to implement.

8

Biotechnology and Waste

As mentioned in Chapter 1, waste represents one of the three key intervention points for the potential use of environmental biotechnology. Moreover, in many ways this particular area of application epitomises much of the whole field, since the management of waste is fundamentally unglamorous, typically funded on a distinctly limited budget and has traditionally been viewed as a necessary inconvenience. However, as the price of customary disposal or treatment options has risen, and ever more stringent legislation been imposed, alternative technologies have become increasingly attractive in the light of their greater relative cost effectiveness. Nowhere has this shift of emphasis been more apparent than in the sphere of biological waste treatment.

With all of environmental biotechnology it is a self-evident truism that whatever is to be treated must be susceptible to biological action and hence the word 'biowaste' has been coined to distinguish the generic forms of organic-origin refuse which meet this criterion, from waste in the wider sense, which does not. This approach also removes much of the confusion which has, historically, dogged the issue, since the material has been variously labelled *putrescible, green, yard, food* or even just *organic* waste, at certain times and by differing authors, over the years. By accepting the single term *biowaste* to cover all such refuse, the difficulties produced by regionally, or nationally, accepted criteria for waste categorisation are largely obviated and the material can be viewed purely in terms of its ease of biodegradability. Hence a more process-based perspective emerges, which is often of considerably greater relevance to the practical concerns of actually utilising biotechnology than a straight-forward consideration of the particular origins of the waste itself.

The Nature of Biowaste

Biowaste arises from a number of human activities, including agriculture, horticulture and industry, broadly falling into one of the following three major categories: faeces/manures, raw plant matter or process waste. This fits neatly into the process-orientated approach mentioned above, since the general characteristics of each are such that biological breakdown proceeds in essentially the same manner within the group and, thus, the ease of their decomposition

Environmental Biotechnology: Theory and Application, Second Edition Gareth M. Evans and Judith C. Furlong
© 2011 John Wiley & Sons, Ltd.

is closely similar. Although, at least chemically speaking, biowaste can be seen as being characterised by a high carbon content, this definition is so wide as to include the vast majority of the substances for which all environmental biotechnologies are viable process options. Hence, in the present discussion, biowaste is limited to substances which have been derived from recently living matter, with the approaches available to deal with other carbon-rich materials having already been examined in the preceding chapters on pollution control, contaminated land and effluent treatment.

Composition of biowaste

Biowaste of animal origin such as that contained in sewage and soiled animal bedding contains unabsorbed fats, proteins and carbohydrates, resulting from incomplete digestion of ingested food of animal and plant origin. In addition, abattoir waste would include all of the above and a substantial proportion of fats and protein, derived from the slaughtered animal. In addition, materials excreted by the animal include metabolic breakdown products such as urea and other small nitrogen containing materials, for example partially degraded bile pigments. Live and dead bacteria, normally resident in animal gut are also present in the biowaste and so contribute their own fats, proteins, carbohydrates and nucleic acids.

In addition to all the components listed above, biowaste of plant origin will contain cellulose, hemicelluloses and lignin. Cellulose is worthy of note given that estimates of over 50% of the total organic carbon in this biosphere is to be found in the form of cellulose. This is unsurprising, since wood is approximately 50% cellulose and cotton is almost 100% cellulose. This macromolecule is an unbranched polysaccharide comprising D-glucose units linked by β 1-4 linkages (see Figure 2.3). It is this β link, rather than the β link found in its animal equivalent, glycogen, which prevents cellulose being broken down by the metabolic pathways in animals. The initial step in the degradation of cellulose is the removal of a glucose molecule from one end of the long chain which is a reaction catalysed by the enzyme *cellulase*. Where cellulose is degraded in animals it is by bacteria resident in the animal rumen or gut which possess *cellulase*. There are also many bacteria living outside the gut, both aerobes and soil anaerobes (Monserrate, Leschine and Canale-Parola, 2001) responsible for cellulose metabolism. Another major constituent of plant material, the hemicelluloses, are also polysaccharides but the subunit in this case is the five carbon sugar D-xylose, also joined 'head to tail' by a β 1-4 linkage. Otherwise, hemicellulose is not related to cellulose despite the similar name. Unlike cellulose which comprises only D-glucose and in an unbranched structure, the family of hemicelluloses has side chains and these may comprise any of a variety of sugars, one of which may be the five carbon sugar, arabinose. The function of hemicelluloses in plants is to form part of the matrix which holds the cellulose fibrils together to improve strength and rigidity of the plant tissue. Lignin is also a very abundant material in plants and is estimated to comprise almost 25% of the dry weight of wood. Totally unlike cellulose or hemicelluloses, which are polymers of sugars and therefore are carbohydrates, lignin is a polymer of the two amino acids, phenylalanine and tyrosine. Despite its abundance, its structure is poorly understood, in part a tribute to the fact that it is extremely resistant to degradation and therefore presents problems to the analyst. Fortunately for the natural process of carbon and nitrogen recycling on which our biosphere depends, fungi degrade lignin and, in addition, some microbes, like those resident in the gut of termites can perform the same function.

Biowaste makes up a huge percentage of refuse; some 2500 million tonnes arises each year in the European Union alone (Lemmes, 1998) and this is a figure which many authorities suggest increases by between 3 and 5% annually. Although the focus of much of this chapter is firmly centred on the biowaste component of Municipal Solid Waste (MSW), since this is the kind of waste which most directly concerns the largest number of people, it is important to be aware that this does not represent the full picture, by any means. Of these 2500 million tonnes of biowaste, 1000 million is agricultural in origin, 550 million tonnes consists of garden and forestry waste, 500 million is sewage and 250 million results from the food processing industry, leaving MSW only to make up the remaining 200 million tonnes. In 2008, the European Commission estimated that the EU as a whole is producing between 75 and 100 million tonnes of biowaste annually from food and garden waste alone. Recent moves to accelerate the production of the long-anticipated draft EU Biowaste Directive – and its eventual implementation – come as little surprise as a result.

The scale of the problem is clearly large, one study suggesting that an annual total of between 850 and 1000 kg (total solids) of material suitable for biological treatment are produced per person (Frostell, 1992). There is general agreement that biowaste accounts for around a third of the industrialised world's municipal waste stream and that a further 30% or so is also expressly biodegradable, such a definition including paper. In the light of this, the fact that the potential for the development and application of approaches based on biological processing has not yet been more rigorously or comprehensively explored remains somewhat surprising. Moreover, with society in general increasingly committed to the 'green' ideals of maximised recycling and the rational utilisation of waste, it is difficult to see how such goals can realistically ever be expected to be met, without significant attention being paid to the biowaste issue. In this respect, the writing may already be on the wall, since the demands of legislation appearing in Europe, the US and elsewhere has already driven fundamental reappraisals of the way in which all refuse is regarded. In particular, regulatory changes designed to reduce the amount of raw biodegradable material destined for landfill must ultimately come to promote biotechnologies which can treat this material in an effective and more environmentally acceptable way. While predicting the future is, of course, notoriously difficult, it seems likely that biological processing will continue to assume an ever-more central role in future waste management regimes, which presents both exciting possibilities and some genuine challenges to the industry itself. However, in order to understand why, it is important to consider the current difficulties posed by biowaste under traditional disposal routes.

Table 8.1   *A comparison of selected national waste management arrangements, recycling rates and MSW biowaste component*

Country	Landfill (%)	Incineration (%)	Recycled (%)	Biowaste (%)
Austria	65	11	24	27
Belgium	43	54	3	47
Canada	67	4	29	34
Denmark	20	55	25	37
Finland	66	4	30	33
France	59	33	8	30
Germany	46	36	18	30
Netherlands	30	42	28	35
Japan	21	74	5	26
Norway	68	18	14	35
Sweden	34	47	19	38
Switzerland	11	47	42	30
UK	85	9	6	30
USA	61	15	24	32

Sources: IEA Bioenergy, European Commission and relevant Embassies.

Although a number of changes in the whole perception of waste have led to a variety of relatively new options receiving attention, generally throughout the world, the vast majority of refuse is dealt with either by means of landfill or incineration. Different countries and administrations have favoured one or the other at various times and, as with all things to do with waste, local custom and circumstance have played a major part in shaping the current status quo. While it is beyond the scope of the present discussion to examine this in any depth, Table 8.1 may help to provide some indication of the wider situation.

Although there has been considerable development in incineration technology over the years and today's facilities, with their energy recovery, power generation and district heating potential, are a far cry from the simple smoking stacks of old, for biological origin waste, mass burn incinerators cannot be viewed as the ideal solution. Hence, while the incineration *versus* landfill argument still rages, and has been re-visited with renewed vigour once again in some circles in the light of the implications of recent European developments, the fact remains that, at least from the standpoint of biowaste, both are nothing more than disposal routes. Significant amounts of wet organic material, which is itself largely composed of water to begin with, may be an inconvenience to the incinerator operator; the situation in landfill is worse.

Landfill

Left to its own devices, all discarded biological waste gradually undergoes a natural process of biodegradation, typically beginning with autolysis and culminating in putrefaction. The speed at which this progresses is governed by a number of factors such as the nature and freshness of the material, the temperature, moisture, and so on. When this happens in the open air, or in the upper levels of the soil, decomposition is aerobic, the organic material being mineralised and carbon dioxide (CO_2) released as the major gaseous product. However, though biowaste awaiting collection in dustbins and even, to some extent, when only recently delivered to landfill, initially begin to breakdown in this way, older putrescible material, buried deeper, experiences conditions effectively starved of oxygen. In this environment, the degradation process is anaerobic and mineralisation continues with broadly equal amounts of methane (CH_4) and carbon dioxide being produced. This resultant mix is known as landfill gas and typically contains a number of trace gases of varying chemical composition. At the functional level, the mechanism of this reaction is very complex, with hundreds of intermediary reactions and products potentially involved and many requiring additional synergistic substances, enzymes or other catalysts. Methanogenesis is discussed more fully elsewhere, but it is possible to simplify the overall process thus:

$$\text{Organic material} \longrightarrow CH_4 + CO_2 + H_2 + NH_3 + H_2S$$

The production of methane is a particular worry in environmental terms since although there is some disagreement as to the exact figure, it is widely accepted as more than 30 times more damaging as a greenhouse gas than a similar amount of carbon dioxide (CO_2). It was precisely because of these concerns that the European Union began its drive to produce statutory controls on the amount of biodegradable material permitted to be disposed of by this route. Without going into lengthy descriptions of the final legislation adopted, or the history of its stormy 10-year passage into European law, it is fair to say that the elements of the Landfill Directive which relate to biowaste require considerable changes to be made in waste management practice. This is of particular importance for those countries, like the UK, with a previously heavy traditional reliance on this method. A series of stepped major reductions in the amount of material entering landfill are required and a timetable has been imposed for their implementation. By 2020 at the latest, all EU Member States must have reduced their biowaste input into landfill by 65% of the comparable figure for 1995. According to the Directive, 'biodegradable' is expressly defined as any 'waste, that is capable of undergoing anaerobic or aerobic decomposition, such as food and garden waste, and *paper and paperboard*' (DETR, 1999a). This has particular implications for currently landfill-dependent nations. Established figures from the Environment Agency show that 32% (by dry weight) of MSW production in the UK is paper. This represents its single largest biodegradable component, using the Directive definition, pushing the traditional biowaste element into second place by 11% (DETR, 1999a). Taking into account the additional contributions of 1% textiles, 3.5% 'fines', 4% miscellaneous combustibles and non-combustibles at 1%, the grand total of 'biodegradable' inclusions in the UK waste stream comes to 62.5%, based on figures from this same study (DETR, 1999a). Making up more than half of the total on its own, paper is, then, of great potential importance, and it is clear that no attempt at reaching the levels of reduction demanded by law can afford to ignore this material.

The question of methane production, so central to the original thrust of the legislation has been addressed by requiring sites to collect the landfill gas produced and use it for energy generation, conceding that it may be flared off where for some reason this is not possible.

A second potential environmental problem typically associated with landfills is the production of polluting leachates, which can be aggravated by the dumping of biowaste. Water percolating through the site tends to leach out both organic and inorganic substances which can lead to contamination of the groundwater. The persistence of pathogens and the potential translocation of many biologically active chemicals have become of increasing concern over the last 10–15 years in the light of growing (though still largely circumstantial) evidence of health problems associated with proximity to certain landfill sites. However, there is considerable variance between many aspects of different facilities and, additionally, much uncertainty as to the extent of any possible exposure to chemicals found therein (Vrijheid, 2000). The UK Government commissioned what arguably remains the world's most extensive study to date into the potential health risks of living within 2 km of landfills, to examine the incidence of low birth weight, congenital defects, stillbirths and cancers in the vicinity of 9565 landfill sites, with a sample size in excess of some 8 million pregnancies. This revealed a 7% increase in the rate of both chromosomal and non-chromosomal birth defects (Elliott *et al.*, 2001) but the expert advisory committee observed that this represented only a small excess risk and might well be accounted for by factors other than those directly attributable to landfill itself. While domestic landfill operations, then, may well be of little significant threat to those around them, the situation for hazardous waste sites, though admittedly less well investigated, appears somewhat different.

The findings of the later investigation (Vrijheid *et al.*, 2002) of data originating from a smaller study of certain European landfills which accept hazardous waste (Dolk *et al.*, 1998) suggests a 40% increase in chromosomal birth defects and a 33% increase in the risk of non-chromosomal abnormalities, within a 3 km radius. However, whether the observed increase in risk arises merely from living near such a hazardous waste site, or as a result of other factors as yet unknown, remained unclear. Subsequently, a 2006 study into the health concerns of residents neighbouring an industrial landfill in Melbourne, Australia failed to identify any difference in the occurrence of cancers, birth defects or other birth-related outcomes between those living in the area and the rest of the state. It did note a small but statistically significant increase in pre-term births (defined as <37 weeks gestation) but since this fitted with a more general regional pattern, socio-demographic influences were cited as the most likely influence (Lynch *et al.*, 2006). Clearly, understanding the true scope of landfill releases, their potential toxicity, the possible exposure pathways and how other environmental or demographic factors inter-react with them to influence health is – and will remain – essential in interpreting any epidemiological data.

Even where there is nothing to suggest an adverse effect on the local population, high concentrations of biowaste-derived leachate remain undesirable. Such rich liquors provide heterotrophic micro-organisms with a ready and abundant source of food. In conditions of relatively low organic loading, a dynamic equilibrium is reached between the bacteria breaking this material down and the autotrophic organisms, typically algae, which subsequently make use of these breakdown products. The oxygen balance works, since the requirements of the aerobic decomposers is offset by the contribution of the photosynthetic algae present. However under conditions of high organic loading, the oxygen demand of the bacteria exceeds the carrying capacity of the water and the algae's ability to replenish it. Hence a downward spiral develops, which ultimate leads to locally anaerobic conditions.

Although 'waste' is itself one of the three key potential intervention points for environmental biotechnology, it should be clear from the preceding discussion that there is considerable capacity for biological waste treatment technologies to contribute heavily to another, namely the reduction of pollution. To try to set this in context, it is quite common for landfill leachate analysis ranges to be quoted based on the average values obtained from a number of established sites. However, this can lead to a significant distortion of the true picture since, particularly for newer landfills (where the biochemical activity tends more to early acetogenic fermentation than 'old' post-methanogenic or even semi-aerobic processes) a degree of under-representation often occurs for some substances. For example, 'young', acetogenic leachate is typically below pH 7 and of high Chemical Oxygen Demand (COD), though much of the latter is biodegradable. The bacteria responsible for the biological breakdown at this point in the site's life may be anaerobic, aerobic or facultative anaerobes. In older landfills, methanogenic bacteria predominate, which are strict anaerobes and can only assume and maintain their dominant position in the absence of oxygen. Such conditions develop in time as the normal sequence of events involves the early acetogenic bacteria gradually using up the available oxygen and producing both the necessary anaerobic environment and acetate as a ready food source for the methanogens which follow in succession, as the site ages.

The full picture of the pollution potential of landfill leachate is more complex than might at first be supposed, if for no other reason than, though it is spoken of as if it were a single commodity, leachate is a highly variable and distinctly heterogeneous substance. It is influenced by the age, contents and management of the landfill of its origin, as well as by the temperature and rainfall of the site. Moreover, all of these factors interact and may vary considerably, even in the relatively short-term, not to mention over the decades of a typical landfill's lifetime. The general range of values for landfill leachate established by the Centre for Environmental Research and Consultancy (CERC) study (Cope, 1995) makes this point very clearly, as shown in Table 8.2.

Some measures have been written into the legislation in an attempt to minimise the possibility of pollution, such as the requirement that all sites, except those taking inert waste, employ a leachate collection system and meet universal minimum liner specifications. However, it is obvious that a method of dealing with waste which removes the bulk of the problem at the outset must be a preferable

Table 8.2 *General range of values for landfill leachate*

Determinand	Value range
pH	5.5–8
BOD	5–80 000
COD	50–100 000
Nitrate	0.1–1 000
Sulphate	1–1 000
VFAs	150–50 000

Concentrations expressed in milligrams/litre, pH in pH units.

solution. The use of biological treatment technologies to process wastes has, then, considerable future potential both in direct application to waste management itself and in a number of allied pollution control issues which currently beset this particular industry. Coupled with the twin external driving forces of legislation and economic forces in the commercial arena, this means that waste biotechnologies seem certain to assume greater importance in the coming decades.

Biological Waste Treatment

The aims of biological treatment are relatively straightforward and can be summed up in the following three points:

1. Reducing the potential for adverse effects to the environment or human health.
2. Reclaiming valuable minerals for reuse.
3. Generating a useful final product.

Broadly speaking, this effectively means the decomposition of the biowaste by microbes to produce a stable, bulk-reduced material, during which process the complex organic molecules originally present are converted into simpler chemicals. This makes them available for literal recycling in a wider biological context.

To some extent these three aims can be seen as forming a natural hierarchy, since removing environmental or health risks, and deriving a stable product, forms the bottom rung of the ladder for all biological waste treatment technologies. Clearly, whatever the final use of the material is to be, it must be safe in both human and ecological terms. The recovery of substances, like nitrogen, potassium and phosphorus, which can be beneficially reused, forms the next level up, and is, in any case, closely linked to stabilisation, because these chemicals, if left untreated within the material, would provide the potential for unwanted microbial activity at a later date. The final stage, the generation of a useful end-product, is obviously dependent on the previous two objectives having been met with some degree of efficiency. The possible uses of the final material, and just as importantly, its acceptability to the market, will largely be governed by the certainty and effectiveness of the preceding processes of stabilisation and reclamation. Thus, while the hierarchical view may, in some ways, be both a natural and a convenient one, these issues are not always as clear-cut, particularly in respect of the implications for commercial biowaste treatment, as this approach might lead one to believe.

In practical terms, the application of this leads to two major environmental benefits. Firstly, and most obviously, the volume of biowaste consigned to landfill is decreased. This in turn brings about the reduction of landfill gas emissions to the atmosphere and thus a lessening of the overall greenhouse gas contribution, while also freeing up space for materials for which landfill genuinely is the most appropriate disposal option. Secondly, good biological treatment results in the generation of a soil amendment product, which potentially can help lessen the demand for peat, reduce the use of artificial fertilisers, improve soil fertility and mitigate the effects of erosion.

As has been mentioned previously, stabilisation is central to the whole of biological waste treatment. This is the key factor in producing a final marketable commodity, since only a consistent and quality product, with guaranteed freedom from weeds and pathogens, will encourage sufficient customer confidence to give it the necessary commercial edge. As a good working definition, stabilisation is biodegradation to the point that the material produced can be stored normally, in piles, heaps or bags, even under wet conditions, without problems being encountered. In similar circumstances, an incompletely stabilised material might well begin to smell, begin renewed microbial activity or attract flies. Defined in this way, stability is somewhat difficult to measure objectively and, as a result, direct respirometry of the specific oxygen uptake rate (SOUR) has steadily gained support as a potential means to quantify it directly. Certainly, it offers a very effective window on microbial activity within the matter being processed, but until the method becomes more widespread and uniform in its application, the true practical value of the approach remains to be seen.

The early successes of biowaste treatment have typically been achieved with the plant matter from domestic, commercial and municipal gardens, often called 'green' or 'yard' wastes. There are many reasons for this. The material is readily biodegradable, and often there is a legal obligation on the householder to dispose of it separately from the general domestic waste. In 1999, the UK production of this type of biowaste alone was estimated at around 5 million tonnes per annum (DETR, 1999b); currently that figure exceeds 7 million tonnes, making this one area in which biological waste treatment can make very swift advances. No where is the point better illustrated than in America, where the upsurge in yard waste processing throughout the 1990s, led to a biowaste recovery rate of more than 40%, which made an effective contribution of nearly 25% to overall US recycling figures. In many respects, however, discussions of waste types and their suitability for treatment are irrelevancies. Legislation tends to be focused on excluding putrescible material from landfills and, thus, generally seeks to make no distinction as to point of origin and applies equally to all forms. The reasons for this are obvious, since to do otherwise would make practical enforcement a nightmare of impossibility. In any case, the way in which waste is collected and its resultant condition on arrival at the treatment plant is of considerably greater influence on its ease of processing and the quality of the derived final product.

There are three general ways in which waste is collected; as mixed MSW, as part of a separate collection scheme, or via civic amenity sites and recycling banks. From a purely biowaste standpoint, mixed waste is far from ideal and requires considerable additional effort to produce a biodegradable fraction suitable for any kind of bio-processing, not least because the risk of cross-contamination is so high. By contrast, suitably designed separate collection schemes can yield a very good biowaste feedstock, as a number of countries around the world have successfully shown. However, not all separate collections are the same, and they may vary greatly as a result of the demands of local waste initiatives and specific targets for recycling. As with all attempts to maximise the rational use of waste, the delivered benefits of any scheme inevitably reflect the overall emphasis of the project itself. Where the major desire is to optimise the recovery of traditional dry recyclables, biowaste may fare poorly. Systems deliberately put in place to divert biodegradable material from landfill or incineration routes, however, generally achieve extremely satisfactory results. In many respects, the same largely holds true for recycling banks and amenity sites. Dependent on local emphasis, the operation can recover very specific, narrow waste types, or larger, more loosely-defined, general groups. Where 'garden' waste is kept separate, and not simply consigned to the overloaded skip labelled 'other wastes', the biowaste fraction produced can, again, be of a very high quality and readily acceptable for biological treatment. Indeed, it is generally accepted that this material is the cleanest source available for processing and it constitutes something in the region of three-quarters of the biowaste treated yearly in the UK (DETR, 1999b).

For those approaches to collection which do not involve separation of the putrescible fraction at source, obviously some form of sorting will be required before the material can be taken on to any kind of biological processing. It lies beyond the remit of this work to attempt to describe the methods by which this can be achieved, or their relative merits. Suffice it to say that whatever on-site sorting is used must be matched adequately to the demands of the incoming waste stream, the intended treatment biotechnology and the available local resources. However, the biowaste-rich fraction is obtained, the major consideration for processing is its physical form, which is of more fundamental significance to biowaste than any other refuse-reclaimed material. For traditional dry recyclables, chipping, crushing or baling are mere matters of convenience; for bio-treatment, particle size, purity and consistency are indivisible from the process itself, since they are defined by the requirements of the microbes responsible. In general terms this means that the biowaste is shredded to break it down into small and relatively uniform pieces, the exact requirements being dictated by the particular treatment technology to be used. This not only makes mixing and homogenisation easier to achieve but also, by increasing the surface area to volume ratio, makes the material more available to microbial action.

There are a number of processes currently available in varying degrees of commercial readiness, and others under development, to deal with biowaste. While the underlying aims and basic requirements of all these biotechnologies are essentially the same, there is some variance of detail between individual methods. Two general approaches in particular, composting and anaerobic digestion (AD), are so well established and between them account for such a large proportion of the biowaste treated worldwide, that the discussion of specific technologies must begin with them.

Composting

For centuries, gardeners and horticulturalists have encouraged biodegradable waste to break down to produce stable, nutrient rich compost for use in pots or directly for improving the soil. This application of the natural, exothermic process of aerobic decomposition, is familiar and time-honoured. More recently, however, composting has been the recipient of increased attention as a potential means of treating biowaste on a municipal basis. Though the scale of such operations imposes certain restrictions of its own, generally, putrescible matter decomposes more efficiently and completely when oxygen is readily available. This leads to proteins being degraded to nitrogen or ammonia and ultimately mineralised to nitrate, while fats and carbohydrates are broken down to carbon dioxide and water, via organic acids. This is, of course, purely a mass flow overview of the process since a proportion of the material becomes incorporated into microbial cells, as the decomposers themselves multiply and grow. Even under optimised environmental conditions, there are a number of rate limiting factors in the process, which include extracellular hydrolytic enzyme production, the speed of hydrolysis itself and the efficiency of oxygen transfer. These may, in turn, be influenced by other aspects such as the particle size and nature of the biowaste material to be treated.

In a practical application, this can be a major consideration as the kinds of biowaste to be composted can vary greatly, particularly when derived from MSW, since seasonal variation, local conditions and climate may produce a highly heterogeneous material. On the other hand, biowastes from food processing or horticulture can be remarkably consistent and homogeneous. Accordingly, the details of breakdown may be very complex, involving a number of intermediary compounds and different organisms utilising various biological pathways. However, in broad terms, the composting process can be split into the following four distinct general phases, which are chiefly defined by their temperature characteristics.

The composting process

- Latent phase: (ambient temperature – circa 22 °C) composting microbes infiltrate, colonise and acclimatise to the material.

- Growth phase: (circa 22–40 °C) growth and reproduction of microbes, resulting in a high respiration rate and consequent elevation of temperature to a mesophilic range.
- Thermophilic phase: (circa 40–60 °C) compost pile achieves peak temperature and maximum pathogen sterilisation. At the end of this phase the temperature drops to around 40 °C.
- Maturation phase: (circa 40 °C – ambient) slower, secondary mesophilic phase, with the temperature gradually dropping to ambient temperature as the microbial activity within the material decreases. Complex organic chemicals are transformed into humic compounds and residual ammonia undergoes nitrification to nitrite and subsequently nitrate.

For a municipal scheme, as shown schematically in Figure 8.1, time and space at the facility will be at a premium, so the faster the biowaste can be colonised by a suitable microbial culture, the sooner the treatment space will be ready to accept a new load for processing. Hence the principal focus of environmental biotechnology for the optimisation of conditions for enhanced biological breakdown is in reducing the time-lag inherent in the latent phase. One of the major means to achieve this is to ensure that the material to be treated is presented in as suitable a form as possible. This typically involves some form of grinding or shredding to produce an ideal physical particle size, but biochemical considerations are every bit as important in this respect. At the same time as breaking down biowaste into simpler compounds, the process also brings about a change in the carbon to nitrogen (C : N) ratio of the material, as substantial quantities of organic carbon are converted to carbon dioxide. The initial C : N ratio is an important factor in the success of composting, since a ratio much more than 25 : 1 can inhibit the mineralisation of nitrogen and adversely affect the product's final maturation. This latter aspect has clear implications for any intended use of the end product as a fertiliser or soil enhancer, particularly for a large scale, commercial operation. To take account of this, facilities accepting mixed-source waste for composting, often find it necessary to undertake a measure of mixing and blending to ensure an appropriate balance. It is possible to categorise different kinds of biowaste according to their carbon/nitrogen content, as illustrated in the Table 8.3.

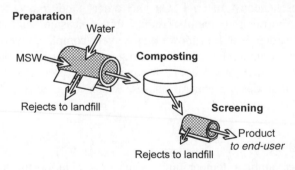

Figure 8.1   Compost plant schematic flow chart

Table 8.3   *Illustrative carbon to nitrogen ratios*

Material	C : N
Food wastes	15 : 1
Sewage sludge (digested)	16 : 1
Grass clippings	19 : 1
Cow manure	20 : 1
Horse manure	25 : 1
Leaves and foliage	60 : 1
Bark	120 : 1
Paper	170 : 1
Wood and sawdust	500 : 1

Most plants obtain the nitrogen they require as nitrate. The mineralisation of nitrogen is performed by two genera of bacteria, *Nitrosomonas*, responsible for converting ammonia to nitrite and *Nitrobacter* which completes the nitrification. Inactivated at temperatures above 40 °C and with a relatively slow growth rate, their activity is largely confined to the maturation phase. Thus, the proper mineralisation of nitrogen can only be achieved after the growth and thermophilic stages have been completed, which themselves require a suitable C : N balance in the initial feedstock in the first place.

A number of different organisms are involved in composting, including bacteria, fungi, protozoa, mites, nematodes, insects and annelids, there being a natural succession of forms allied with the four phases of the process. Thus the initial decomposition is brought about by mesophilic bacteria in the main, until their increased activity raises the temperature into the range favoured by thermophilic organisms. These thermophiles then play a major role in the breakdown of carbohydrates and proteins, before they themselves become inhibited by the 70–75 °C heat of the composting pile. Later, as the temperature begins to drop, actinomycetes become the dominant group, giving the ageing compost a characteristic white-grey appearance. Although largely confined to the surface layers, they are of considerable importance in the decomposition of cellulose and lignin, which are two of the more difficult components of biowaste to break down.

The microbial component of compost is an area of particular future potential, particularly as a measure of product quality. While simple chemical analysis has traditionally been used to assess composts, concentrating on NPK values and placing it on the same footing as artificial fertiliser, there has been a growing realisation that its complex nature means that this does not tell the full story. The potential benefits in terms of soil flora improvement and plant pathogen suppression cannot be inferred from a compost's gross mineral contribution and so, in an attempt to produce a more comprehensive yardstick, some producers and users have begun to investigate assessment based on microbiological profiling. Pioneering work in the US, by BBC Laboratories of Arizona, has led to the development of the first predictive tool for the value of a compost as a soil microbial inoculant, based on the concentrations of six key classes of micro-organisms present (Bess, 1999). The marketing of biowaste derived products is made more difficult by a number of factors which lie outside of the scope of this work, most particularly a lack of a recognised and universally agreed standard. The application of microbiological criteria, in conjunction with the likes of mineral analysis and maturity assessment, could lead to significantly better overall characterisation of composts. Moreover, there seems no reason why this approach should not be extended to all biowaste-derived soil amendments, enabling direct comparisons to be made and the suitability of any given product for a particular use to be measured objectively.

Applying Composting to Waste Management

Composting has an appeal to local authorities needing to meet diversion targets while keeping a watch on their budgets, since it is relatively simple and does not demand particularly high resource investment, either to set up or run. As a consequence, many of the initiatives instigated to deal with biowaste have been based on composting of one form or another. In the broadest of terms, such schemes fall into one of two categories, namely, home composting, or centralised facilities. The focus of this section will fall on the latter, as a more representative application of biotechnology, though to set this in context, it is worth giving a brief outline of the former.

Home composting

Home based systems differ little in reality from the traditional gardener's approach, putting biodegradable material into a heap or, more typically a bin, often provided free or at a subsidised price, by the local council. Though this does have the advantage of directly involving people in the disposal of their own waste and the informality of this approach has its own advantages, such schemes are not without certain drawbacks. To work, these initiatives draw heavily on householder goodwill and competence, not to mention a good choice of bin and simply making the means available does not, of course, ensure that it will be used. Anecdotal evidence suggests that many bins lie unused within two years, once the initial enthusiasm wears off, and an investigation into Luton's trial scheme suggested that home composting may make little difference to the overall amount of waste generated (Wright, 1998). The kind of instant minimisation popularly supposed would seem to be far from guaranteed.

One clear advantage that household composters do have, however, is the ability to control very closely what goes into their system. This avoids both the issues of contamination and the need for post-user segregation typically foisted on the operators of centralised facilities. Thus, although domestic initiatives of this kind are unlikely ever to make the sort of difference to biowaste treatment demanded by legislation on their own, it seems likely that they will always have a role to play, perhaps most especially in remoter areas where collection for processing elsewhere might prove uneconomic.

Centralised composting

The biochemistry and microbiology of all composting remains essentially the same, irrespective of the details of the operation. However, the scale of schemes set up to deal with a municipal biowaste stream in terms of the physical volume involved imposes certain additional considerations, not least amongst them being the need to ensure adequate aeration. In the back-garden compost heap, oxygen diffuses directly into the material; large-scale composting cannot rely on this method, as the large quantities involved lead to a lower surface area to volume ratio, limiting natural oxygen ingress. To overcome this various techniques make use of mixing, turning or pumping, but, clearly, the additional energy required has its own implications for a commercial operation.

Approaches suitable for municipal scale use fall into five main categories:

- windrow;
- static pile;
- tunnel;
- rotary drum;
- in-vessel.

A sixth form, termed 'tower composting', may occasionally be encountered, but it is generally much less common than the other five.

No one system is the universal ideal. The decision as to which approach is likely to be the most suitable for specific conditions is dependent on a number of factors, including the nature and quantity of biowaste available, the required quality of the end product, the time available for processing, local workforce and land availability and financial considerations. In the brief descriptions of each system which follow, the main features are set out, but it is important to realise that there are many other issues which have a bearing on the selection of appropriate technologies than the necessary brevity of this chapter permits us to explore fully.

Windrow

The biowaste is laid in parallel long rows, around 2 or 3 m high and 3 or 4 m across at the base, forming a characteristically trapezoid shape. Windrowing is usually done on a large scale and, though they can be situated under cover, generally they tend to be outdoor facilities, which exposes them more to the vagaries of the weather and makes process control more difficult. While this might be a problem for some kinds of biowaste, for the typical park and garden waste treated by this method, it generally is not. However, some early attempts were prone to heavy leachate production in conditions of high rainfall, leading to concerns regarding localised soil pollution. This was largely an engineering problem, however, and the almost universal requirement for a suitably constructed concrete pad and interceptor has made this virtually unknown today.

Limited aeration occurs naturally via diffusion and convection currents, but this is heavily augmented by a regime of regular turning, which also helps to mix the composting material, thus helping to make the rate of breakdown more uniform. Dependent on the size of the operation, this may be done by anything from front-end loaders on very small sites, to self-propelled specialised turners which straddle the windrows at larger facilities. The intervals between turning can be tailored to the stage of the process, being more frequent early on, when oxygen demand is high, becoming longer as composting proceeds.

Windrows have a typically high land requirement, can potentially give rise to odour problems and are potentially likely to release fungal spores and other bioaerosols during turning. Despite these drawbacks, this approach accounts for the vast majority of centralised composting projects, possibly because it is often carried out as an addition to existing landfill operations, thereby significantly reducing the actual nuisance generated.

Static pile

Superficially resembling the previous method, the static pile, as its name suggests, is not turned and thus does not have to conform to the dimensions of a turner, allowing the rows to be considerably taller and wider. What mixing is needed can be achieved using standard agricultural equipment and so these systems tend to be significantly cheaper in respect of equipment, manpower and running costs. They do not, however, remove the land requirement, since decomposition progresses at a slower rate, causing the material to remain on site for a longer period.

In an attempt to get around this, a variant on the idea has been developed, particularly for the co-composting of food or garden biowaste with manure or sewage sludge, which relies on forced aeration. With a perforated floor and fans to push air through the material, the characteristically low oxygen level within the core of traditional static piles is avoided and processing acceler-ated. However, bulk air movement is expensive, so this system tends to be reserved for small tonnage facilities, often in areas where good odour control is of major importance.

Tunnel composting

Tunnel composting has been used by the mushroom industry for a number of years, where processing takes place inside closed tunnels, around 5 m high and up to 40 feet in length. There has been some interest in adapting it to deal with MSW-derived material and one system which has evolved uses huge polythene bags, a metre or so high and 60-m long into which a special filling machine packs around 74 tonnes of source separated putrescible material. This particular design also makes use of a fan to force air through the material rather like the previous technique, with slits in the side wall allowing carbon dioxide to escape.

The processing time is reduced compared with a similar sized aerated static pile, since the environmental conditions within tunnel is easier to control, though similar cost considerations apply.

Rotary drum

Rotary drums seem to drift in and out of fashion, often being favoured by those needing to co-compost sewage sludge with more fibrous material, like crop residues, straw or garden waste. The principle is simple; the waste is loaded into the drum which then slowly rotates. This gently tumbles the material, mixing it and helping to aerate it. The drums themselves are usually steel, insulated to reduce heat loss.

In-vessel

Sometimes also called Closed Reactor Composters, there are a number of designs of in-vessel systems available, ranging from small steel or plastic tanks, through larger metal cages to long concrete troughs with high sidewalls. The main characteristic of these systems is that the waste breaks down within an enclosed container, which allows the internal environmental conditions to be closely controlled. This approach offers a very efficient use of space and close regulation of the process, since some form of mechanical aeration is also required it is significantly more expensive on a tonne for tonne basis than the less resource-intensive methods. Accordingly, it is less suitable for large capacity requirements, but has a role in smaller scale operations or where the material to be treated does not easily fit into other kinds of processing or disposal arrangements.

This is less of a natural group than the preceding approaches to composting, since it encompasses far greater variety of design. Consequently there is a marked variance in the capacity, complexity and cost of these systems.

Process parameters

Aside of aeration, which has already been discussed, a number of other parameters affect the composting process. Although these are themselves influenced to some extent by the method being used, in general the most important of these factors are:

- temperature;
- moisture content;
- particle size;
- nature of the feedstock;
- accelerants;
- processing time.

Temperature

The temperature changes over the stages of composting have important implications for the efficiency of the process. It is widely agreed that for satisfactory sanitisation, the material should reach at least 55 °C, though there is less of a consensus over the required duration of this exposure.

On the other hand, the temperature should not be permitted to exceed 70 °C, since above this most of the compost microbes either die off or become inactivated, causing the biological breakdown to slow or stop. In a commercial operation, lost processing time has inevitable financial consequences.

Moisture content

A moisture content of around 60% is the ideal target for optimum composting, though anything within a range of between 40 and 70% will suffice. While some biowastes meet this requirement naturally; others forms can be surprisingly dry, sometimes with a moisture content as low as 25–30%, which approaches the levels at which severe biological inhibition can occur. Equally, too wet a material may be a problem as this may restrict aeration and even encourage leaching. Even when the initial mix is right, composting matter gradually loses moisture over time and evaporative losses from the surface of the composting biowaste can cause problems, especially in frequently turned windrow regimes. Careful monitoring and appropriate management is necessary to ensure that the optimum range is maintained.

Particle size

The optimum particle size for composting is, of necessity, something of a compromise. The smaller the individual pieces, the larger the surface area to volume ratio, which makes more of the material available to microbial attack, thus speeding up the process of decomposition. However, particles which are shredded too finely will tend to become compacted and so reduce aeration within the material. Consequently, a balance must be struck, providing the smallest possible particle size which does not interfere with air flow. Individual design features may need to be considered; dependent on the system used, bed depth, aeration method and the nature of the biowaste itself can all have an influence.

Nature of the feedstock

The importance of the carbon to nitrogen (C : N) ratio and the need for careful management to ensure a proper balance has already been discussed. In addition, for some materials, the use of amendments or co-composting with other wastes can also help optimise conditions for biological treatment. Sewage sludge and manures are often used in this way, but they can also boost the available nutrient levels, often in an uncertain way and by variable amounts. Generally, additives are used where there is a need to improve either the chemical or the physical nature of the composting material. Clearly, for a large commercial operation, it is essential that whatever is used does not significantly affect the economics of the plant and for this reason, although artificial fertilisers are an ideal way of increasing nutrient content, they are seldom used in household waste applications. Their expense relative to these low-cost biowastes effectively rules them out; they are, however, often to be seen in *ex situ* bioremediation operations, since composting contaminated soil commands a higher price.

Additions to the original material typically accelerate processing, but careful monitoring is essential since the blend may exhibit very different decompositional characteristics, which may ultimately influence the nature of final product derived.

Accelerants

Although gardeners have a number of proprietary brands of compost accelerants available to them, this is not an approach often used at commercial facilities, mainly due to the scale of these operations and the consequent expense. As with nutrient addition, this tends to be reserved for use on high value wastes, though many common substances used in co-composting programmes, like manures, are themselves widely accepted to act as natural accelerants. Though their effect is more variable, it seems likely that this is the only form of enhanced processing applicable for general biowaste use.

Processing time

In many respects, the time required is a function of all the other factors. Processing garden or food waste can be achieved in under three months using aerated, in-vessel or turned windrow systems, while in a simple static pile, it may take a year or more to reach the same state. Inevitably much depends also on the management regime, since process optimisation is the key to accelerated bio-treatment and good operation practice is, consequently of considerable importance.

Anaerobic Digestion

Although composting certainly accounts for the majority of biowaste treatment applications around the world, AD is an alternative option which has continued to receive increasing interest over recent years. In many respects, it is a regulated version of the natural events of landfill, in that it results in the controlled release of methane-rich biogas, which offers the potential for a very real form of energy-from-waste. This technology is viewed in certain circles as rather novel, but this is not really the case. It has been used in the water industry for around a 100 years to treat sewage and, more recently, been successfully applied to the processing of agricultural and household wastes, most notably in Germany and the Netherlands. However, waste management tends to be a naturally cautious field and the relative lack of a proven track record with MSW derived biowaste compared to composting has made the uptake of this approach slow.

The key to effective practical applications of AD technology lies in regulating and optimising the internal environment of an enclosed bioreactor vessel such that the ideal conditions for the process are produced and maintained. Under these circumstances, in the absence of free oxygen, anaerobic bacteria convert the large organic molecules mainly into methane CH_4 and carbon dioxide CO_2.

The actual progression of this breakdown is chemically very complex, potentially involving hundreds of intermediary reactions and compounds, many of which have their own additional requirements in terms of catalysts, enzymes or synergistic chemicals. Unlike composting, AD occurs at one of three distinct temperature ranges, namely:

- Cryophilic ($<20\,^\circ$C).
- Mesophilic ($20-45\,^\circ$C).
- Thermophilic ($>45\,^\circ$C).

Since AD is very much less exothermic than composting, within a landfill or in bogs and swamps, it proceeds under cryophilic conditions. This largely accounts for the relatively protracted timescale and the irregular progress of breakdown typically encountered in these examples. In order to overcome these drawbacks, engineered anaerobic bioreactors are usually run at one or other of the higher ranges, with additional heat supplied by external means to elevate the temperature to the required level. A variety of technology vendors have developed commercial systems based around either thermophilic or mesophilic digestion, which have their own particular characteristics. Without entering into a lengthy discussion of the relative merits of these approaches, it is important to note that the internal conditions favour different bacterial complements and that certain aspects of the reaction details also differ. Consequently, for any given application, one or other may be particularly suited, dependent on the specifics of the material to be processed and the overall requirements for treatment.

The digestion process

Hydrolysis

Carbohydrates, cellulose, proteins and fats are broken down and liquefied by the extracellular enzymes produced by hydrolytic bacteria. The proteins are broken down into amino acids, fats into long-chain fatty acids and carbohydrates into simple sugars, while the liquefaction of complex biological polymers, especially cellulose, to simple, soluble substances is often the rate limiting step in digestion. The rate of hydrolysis is governed by the nature and availability of the substrate, bacterial population, temperature and pH.

Acidogenesis

The monomers released by hydrolysis, together with volatile fatty acids (VFAs) derived from the protein, fat and carbohydrate components of the material being treated are converted to acetic, lactic and proprionic acids, the pH falling as the concentrations of these rise. Methanol and other simple alcohols, carbon dioxide and hydrogen are also produced during acidogenesis, the exact proportions of the different by products being dependent on bacterial species and the environmental conditions within the reactor. Though we have considered these events as a single stage in the process, some authorities prefer to sub-divide them into *acidogenesis* and *acetogenesis* to highlight the importance of acetic acid, which accounts for around 75% of the methane produced by the next step.

Methanogenesis

Relying on obligate anaerobes whose overall growth rate is slower than those responsible for the preceding stages, this final phase involves the production of methane from the raw materials previously generated. Of these, acetic acid and the closely related acetate are the most important, for the reason mentioned above. There are other potential substrates for methanogenic bacteria, but we will limit the scope of this discussion to the most usual examples, as represented in the following equations:

Acetic Acid:

$$CH_3COOH \longrightarrow CH_4 + CO_2$$

Methanol:

$$CH_3OH + H_2 \longrightarrow CH_4 + H_2O$$

Carbon dioxide and hydrogen:

$$CO_2 + 4H_2 \longrightarrow CH_4 + 2H_2O$$

Methanogenic bacteria also play an important part in the wider overall break-down process, since by converting VFAs into methane, they effectively act to limit pH decrease within the digester. With the acid/base equilibrium naturally regulated in this way, any potential bacterial inhibition by acidification is effectively overcome. This is particularly important for methanogens themselves, since they thrive in a relatively narrow pH threshold of 6.6–7.0, becoming progressively more impaired as the pH falls below 6.4. In this event, the persistence of unmodified VFAs can have potentially serious implications for the final use or disposal of the material derived.

There are four main groups of bacteria involved in AD, as shown below, with some typical examples of each:

- Hydrolytic fermentative bacteria: *Clostridium* and *Peptococcus*.
- Acetogenic bacteria: *Syntrophobacter* and *Syntrophomonas*.
- Acetoclastic methanogens: *Methanosarcina* and *Methanothrix*.
- Hydrogenotrophic methanogens: *Methanobacterium* and *Methanobrevibacterium*.

In reality, these are not the only species present in a digester and, though the stages previously described represent the main desired biochemical reactions, a number of additional bacterial types and biochemical pathways play a role in the overall breakdown process. As with composting, there is much interaction between these various organisms.

Applying AD to Waste Management

The nature of AD inevitably means that its applications to waste management are relatively large scale operations, there being no effective equivalent of home composting. Hence, whether the application is as an on-site treatment for process effluent or as part of a centralised municipal waste initiative, the approach relies heavily on engineering, a schematic plant being shown in Figure 8.2. This is in clear contrast with composting and, together with the attendant additional costs, probably goes further to explain the overall lower take up of this technology than any other factor. It should also be apparent that more resources, and primarily a more skilled workforce, are essential prerequisites for success. However, for wastes which are particularly suited to this form of biotechnology, a number of cost benefit analyses over the years have shown that these drawbacks may often be outweighed by the advantages inherent in the system. As with so many practical applications of environmental biotechnology, there is seldom one catch-all solution and the most appropriate approach can only really be judged on the specifics of the problem. There will always be cases when either composting or AD is self-evidently the most suitable route; when the matter is less clear-cut, however, the technology decision is often much more difficult to make.

There are many ways in which AD systems may be categorised as will be briefly discussed below. However, it is important to realise that, irrespective of their individual construction, they all fundamentally consist of isolated vessels of some kind, designed to exclude air and maintain internal conditions at the optimum for bacterial action. It is possible to describe systems treating slurries of 15% total dry solids (TDSs) or less as 'wet', or 'dry' if their TDS exceeds this

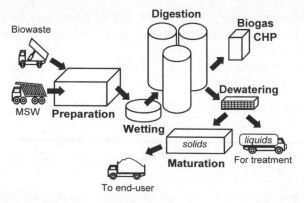

Figure 8.2   *AD plant schematic flow chart*

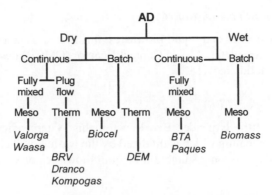

Figure 8.3   *Classifying anaerobic digestion systems*

figure. Alternatively, the temperature range at which they are operated can be used, thus leading to defining AD systems as either mesophilic or thermophilic.

The loading regime adopted can also be a useful means of distinguishing digester types for some purposes, allowing a distinction to be drawn between 'batch' and 'continuous' systems. The former are filled in a single go, then permitted to digest the contents before being emptied and recharged, while the latter have a continuous cycle of new biowaste being added and processed material being drawn off. Figure 8.3 shows the inter-relationship of these various branches of the AD family tree.

However, these are, in effect, operational criteria and as such, though useful in themselves, they can tend to unite dissimilar technologies within essentially artificial groupings, giving little clue as to which best suits what type of biowaste. For this, an examination of aspects of the digester design and engineering principles can often provide a better insight, as in the following descriptions of some of the major examples.

Anaerobic baffled reactor (ABR)

These generally feature a horizontal flow of biowaste through the digester vessel, and are suitable for a wide range of materials. The Valorga process, with its patented gas recirculation and mixing system, is based on this approach.

Anaerobic fixed film reactor (AFFR)

These digesters have a fixed growth plate on which a bacterial biofilm is established, digestion taking place on this surface. They are ideal for relatively weak biowastes with low solids content, but are of less use in other applications.

Completely mixed contact reactor (CMCR)

In this design, the biomass derived during processing is recycled after dewatering to increase the retention time for the solids. This approach is typically used to treat high strength, industrial biowastes.

Continuously stirred tank reactor (CSTR)

Intended to treat slurries and liquid biowastes, this system is essentially the same as the preceding Completely Mixed Contact Reactor (CMCR) design, but without the need for the solids recycle.

Fluidized bed reactor (FBR)

Sometimes termed the Expanded Bed Reactor, this system relies on an internal microbial growth medium which is fluidized by the waste liquid circulating within it. Accordingly they are only suitable for liquid biowastes or very dilute slurries.

Multi-phasic processes (MPPs)

Physically separating the stages of AD into different reactors, these systems are principally of use as experimental tools for increasing understanding of the pathways and mechanisms of the AD process. They are featured here for completeness, but they do not generally have any commercial application.

Upflow anaerobic sludge blanket (UASB)

These systems hold relatively high numbers of active bacteria, which makes them suitable for treating biowastes of low solids content and they are commonly used to process high strength industrial waste liquids or light suspensions.

From all of this discussion, it should be obvious that, however proprietary AD technologies are classified, no one type is universally ideal or superior, each having certain characteristics which make it appropriate for particular wastes and less suitable for others. This means, of course, that although comparisons of the various approaches are of great interest to potential users, in practice they are difficult to make in any meaningful way. Even certified data from an operating plant can only be taken as broadly indicative of how a similar one might perform elsewhere, especially in respect of breakdown efficiency and biogas generation or quality.

Process parameters

Efficient AD requires the development and maintenance of an optimised internal environment to facilitate biological activity. This is of particular importance in the commercial setting and a number of both physical and chemical factors must be taken into account to achieve it, of which the most important are:

- temperature;
- retention period;
- agitation;
- wetness;
- feedstock;
- loading rate;
- pH and VFAs.

Temperature

As mentioned previously, in commercial systems, digesters are operated at around 35 °C (mesophilic) or 55 °C (thermophilic). Irrespective of which approach is adopted for any particular application, a relatively constant temperature is essential for the process to run at its greatest efficiency.

Retention period

Although the amount of biowaste degraded depends on its character, the availability of bacteria and the time allowed for processing, temperature governs both the rate of breakdown itself and the particular bacterial species present in the digester. Hence, there is a direct relationship between temperature and the retention period. Some AD technologies have attempted to shorten the retention period by separating the stages of the process within the digester. The separation of the acidogenic and methanogenic stages permits each to be optimised and this has been well demonstrated at laboratory scale using a completely mixed digester, with phase-isolation being achieved by pH manipulation. Despite the greater efficiency, higher biogas yield and enhanced process stability claimed, it has seen little large-scale use, probably as a result of the higher cost implications of such a system.

Agitation

The agitation of the digester contents has a number of benefits, one of the most obvious being that it helps to mix up material, evening out any localised concentrations, thus also helping to stop the formation of 'dead zones' or scum. In addition, it increases the waste's availability to the bacteria, helps remove and disperse metabolic products and also acts to ensure a more uniform temperature within the digester. There have been some suggestions that efficient mixing enhances methane production, but the evidence is inconclusive, so it seems likely that this may only be of noticeable benefit for some systems or operational regimes.

Wetness

AD is a wet process and any biowaste which is too dry in its natural state will require the addition of a suitable liquid, typically water, recycled AD process liquor or slurries, either sewage or agricultural, before processing can begin. In order to minimise digester size, so-called 'dry' systems have tended to dominate the commercial world, but the relatively thicker contents inevitably demand more energy to mix effectively, off-setting much of the advantage. Comparisons of 'wet' or 'dry' approaches, like those of mesophilic or thermophilic processes, generally yield no clear winner. Each system has particular advantages and applications for certain kinds of biowaste, and selecting the right one for any given use is almost always best done on this basis.

Feedstock

As with composting, the particle size and nature of the material to be treated plays an important role. The ease of breakdown is largely defined by the characteristics of the biowaste material to be treated, but generally finer particles allow for better processing and a homogeneous slurry or suspension is the ideal feedstock for AD. It must be stressed, however, that some biowaste types, particularly the likes of lignin-rich, woody material, are relatively resistant to this process.

Loading rate

Loading depends on the characteristics of the waste, its degree of wetness, digester volume, the expected retention period and similar system design parameters. It is typically expressed as the COD per cubic metre of digester void-space (COD/m^3) or, for continuous or semi-continuous process, per unit time, (COD/day, COD/h).

pH and volatile fatty acids concentration

These are interdependent factors which need to be considered together. Adequate process control and digester optimisation requires suitable pH monitoring, since many of the bacteria involved are pH sensitive. In particular, acidogens, having better tolerance to acidity, may produce acids faster than the increasingly inhibited methanogens can use it, in conditions of low pH, leading to spiralling acidity and the potential for process collapse. A number of acid-base reactions exist within the typical AD process, which lead to a measure of natural, in-built resistance to major pH swings. However, under certain circumstances, the need for external interference may arise and the amount of such intervention necessary to maintain proper equilibrium will depend on the nature of the material. For some wastes, pH control may only be necessary during start up or in overload conditions; for others where acidity is habitually shown to be a problem, continuous control may be necessary.

VFA concentration is one of the most important process indicators. Elevated VFAs are characteristic of AD instability and thus they may be the first indication of a developing problem, though the actual cause may be less immediately obvious. Inadequate mixing, excessive loading, poor temperature control or bacterial inhibition can all lead to an increase in VFAs and a decrease in pH. Considering the inconvenience and cost of being forced to empty a sick reactor, commercial AD operations rely greatly on routine monitoring of this kind.

Biogas

The production of methane-rich biogas, which is an essential characteristic of AD, has been used as an important marketing tool, since the generation of energy from waste by biological means has much attraction, not least amongst those countries with a popular dislike of incineration. A cohesive discussion of bioenergy appears in a later chapter and accordingly, a consideration of the wider implications of biogas fuel will be deferred until then. However, it is worth pointing out in the waste-specific context that, though much has been made of the apparent dual benefit of biowaste treatment *and* energy production, in an actual application, it is a practical impossibility to optimise both simultaneously.

Other Biotechnologies

Although composting and AD between them account for by far the bulk of biological waste treatment globally and each has a well established track record, as is so often the case with practical applications of environmental biotechnology, neither is a clear winner. Dependent on the specifics of the situation, the particular waste composition, local factors, and so on, either can have clear benefits to offer and, as we have seen, both can form adequate responses to the growing demand for biowaste diversion. At the same time, each has its limitations also. For AD, the air-tight nature of the reactor, the gas handling arrangements needed to guarantee that the potentially explosive methane produced can be safely managed and the demands of internal environmental control contribute heavily to the overall capital cost of the plant. In addition, on a practical note, there are certain inherent limits on the levels of contamination by other waste fractions that can be tolerated. Clearly, for some applications, these factors may prove major barriers to use.

On the other hand, composting is essentially less of an engineered solution and in many of the versions often seen at local authority sites, it is a very simple process. The major practical limitation, at least as a sole method of bulk treatment, lies in the physical amount of material, since the typical retention period for composting is longer than AD and the final volume of product derived is greater. Consequently, a relatively large area of land is required for processing and a sizeable market capacity or disposal arrangement is necessary for the compost. In certain circumstances, these issues may be significant blocks to its adoption.

Applications for which neither of the major technologies is an immediately easy fit have, at times, led to interest in the potential of other methods of biowaste treatment. It is beyond the scope of this book to discuss the wider political and economic issues surrounding biowaste initiatives, though it should be clear that such factors commonly play an indivisible part in their implementation. Suffice it to say that these local modalities can often form the most critical deciding factors in determining the suitability of a given approach. This is something of a mixed blessing, since, though it can make direct comparisons between individual methods exceedingly difficult to do in a meaningful way, it does leave space for novel or less-well-known technologies to play a role.

Annelidic conversion (AC)

The use of a variety of annelid worm species is one alternative approach that has received fairly regular re-awakenings of interest over the years, having been variously termed *worm composting*, *vermicomposting*, *vermiculture* or our pre-ferred *annelidic conversion* (AC), a term first attributed to H. Carl Klauck of Newgate, Ontario. The description *worm composting* and its like is somewhat misleading, since the process from both biological and operational criteria is quite distinct from true compost production in two significant ways.

Firstly, as we have seen, in traditional composting, breakdown is brought about by the direct actions of a thriving microbial community. Within a worm-based system, while micro-organisms may contribute in some way to the overall biodegradation, their role in this respect is very much incidental to that of the worms themselves.

Secondly, in worm systems, biowaste is typically laid in much shallower layers than is the case for windrows or static piles, frequently being deposited on the surface of an underlying soil bed. This is a major difference, principally because it reduces the natural self-heating tendency within the decomposing matrix.

Worms of various species can be present in traditional compost heaps, even in thermophilic piles, but they avoid the genuinely thermophilic core, being found at the significantly cooler edges of the heap. In addition, under such conditions the resident annelid population is, in any case, many magnitudes smaller than in the deliberately high-biomass levels of AC systems. While in common with all poikilothermic organisms, worms do require some warmth to remain active, which for most species means a lower limit of $10\,^{\circ}C$, they do not generally tolerate temperatures in excess of $30\,^{\circ}C$ and death occurs above $35\,^{\circ}C$. Most species have an optimum range of $18-25\,^{\circ}C$, which makes the point very clearly that the highly exothermic conditions encountered as part of the 'true' compost-ing process would be impossible for them to survive, and certainly not in any sizeable numbers.

Annelidic conversion is similar to composting in the sense that it can be scaled to meet particular needs and, as a result, it has been promoted in various forms for both domestic and municipal applications over the years. Again, like composting, particularly in respect of home bins, this has not been entirely free of problems, since all the difficulties regarding bin design, operator diligence and issues of compliance apply if anything, more rigorously to AC as to traditional composting. While some recycling officers have found that these projects have been widely welcomed and effective, others report 'considerable' drop-off rate in usage.

In the case of commercial scale treatment, worm systems have sufficient in-built flexibility to be tailored to suit. However, since the beds must be significantly shallower than an equivalent windrow, accommodating the same amount of mate-rial for treatment necessitates a much larger land requirement, which may itself prove a constraining factor. Thus, for each tonne of biowaste to be deposited weekly, the typical bed area required is around $45\,m^2$. Hence, for a typical Civic Amenity Site annual production of 4000 tonnes, and allowing for the seasonal nature of its arising, around half a hectare, or one and a quarter acres, of ground is required simply for the beds themselves. This rises to more than double to provide the necessary service access between and around the wormeries.

Worm systems are essentially biomass intensive, with an initial population density, typically exceeding 500 animals per square metre and a cumulative annelid biomass production rate, once established, of $0.07\,kg/m^2$. Clearly, this demands careful control of the local environmental conditions within the beds for optimisation of system performance, particularly since the physical and biological needs of the organisms involved lie within more precisely defined limits than those of the microbes responsible for composting. Bed design is partly influenced by the temperature tolerances discussed previously, but the large surface area to volume ratio typical of this method also allows for the ready aeration of the biowaste matrix, especially in the surface layers, where many of the worm species used preferentially reside. Design is further constrained by the need for adequate moisture to permit gas-exchange across the annelid skin, which must be achieved without water-logging, since this reduces pile aeration and may further actively drive the resident worms to leave in search of drier conditions. Unsurprisingly, many commercial scale systems make use of extensive drainage works to help avoid this. Well-ventilated covers are also often used, particularly in outdoor installations, which help overcome the rigours of the weather while also producing continuously dark conditions. Accordingly, the worms are encouraged to be active for much more of the day than would otherwise be the case. This brings additional bonuses, since the burrowing of the animals themselves both promotes enhanced aeration and has beneficial effects on odour control, particularly in respect of sulphide concentrations, which have been reported as being reduced by a factor of a 100 or more.

In common with trends in many other biotechnological interventions, there has been some interest over the years in developing in-vessel systems. This principally arises as an attempt to create circumstances in which process control can be maximised, but has the additional benefit of also giving rise to a modular and highly portable approach, which has helped annelidic conversion penetrate areas that might otherwise have remained closed to it. One of the reasons for this is that this approach gets around the need for a permanent installation, which may be an important consideration for some applications. However, the unit processing cost is consequently higher than would be the case for a simple land-based system of similar operational capacity.

A number of different species of worms are used in vermicultural operations around the world, but in general terms all of them can be placed into one of two broad categories, namely redworms and earthworms. Although some uncertainty exists as to the absolute validity of this division, it is a useful tool, at least at the functional or morphological level, to aid understanding of the whole approach. The true earthworms are burrowers, and generally speaking consume dead biological material from within the soil itself rather than directly assimilating the biowaste. The nutrient value of the organics so treated is returned via worm-casts. Hence, initiatives reliant on earthworms are probably best regarded as a form of worm-enhanced composting. By contrast, redworms, which are also sometimes termed *manure* or *compost* worms, rapidly and directly feed on the biowaste, consuming half or more of their own body weight per day. This influx of material is turned into increased worm biomass, both in terms of individual growth and population increase. As a result, AC has tended to make use of redworm species such as *Dendrobaena, Helodrilus* and especially *Eisenia* as the mainstay of these operations. In nature, these animals are naturally found amongst the fallen plant material of woods and forests, where they are commonly associated with the production of leaf litter. Their use in worm beds stands as another example of making use of an organism's natural abilities to achieve the biotechnologist's desired result. When the artificial environment of the worm bed is well enough managed and conditions suitably optimised, the redworms decompose and mineralise the biowaste extremely efficiently, effectively reprising their role in nature under these engineered conditions.

Despite the predominance of redworms within vermiculture for the reasons outlined, there have been a number of cases that have used true earthworm species, with varying degrees of success. Thus, members of the genera *Lumbricus* and *Amynthas* have featured, but perhaps one of the most notable successes was with *Pheretima elongata*, a deep burrowing worm, native to India, which was used to great effect as part of the Bombay plague prevention project, which was founded in 1994 in response to an earlier outbreak of the disease. The growing amount of waste in the city had been strongly implicated in attracting and harbouring the vector rats. The use of vermiculture proved very effective in reducing the biowaste problem (Menon, 1994) and hence was seen as a major preventative measure against a recurrence of the plague.

The use of worms to break down waste is not a particularly novel idea, though like many aspects of environmental biotechnology, it frequently seems to fall first in and then out of favour on a fairly regular basis. The advantages of worm-based biowaste treatment are fairly easy to see. In the first instance, vermiculture offers a high potential volumetric reduction, often exceeding 70% while producing a well-stabilised final product. Secondly, this product itself is rich in potassium, nitrogen, phosphorous and other minerals, which is presented in an ideal form for plant uptake, and hence represents a high fertiliser value. The market for this product has already been successfully established in some parts of the world and many others seem set to follow. This is an important issue, as it has been estimated that for every 1 tonne of biowaste deposited on the bed, around half a tonne of worm casts is produced (Denham, 1996).

Finally, the rapid growth potential of worms under idealised conditions has the potential to provide a harvestable biomass resource, typically either as seed populations for other vermiculture operations or for direct sale into the fishing market. The reputation of the bait outlet has been tarnished in the light of various spectacular collapses, most notably the Californian pyramid franchise back in the 1970s and other similar operations in the UK more recently. Often aimed at farmers seeking diversification ventures, it is unlikely that worm production will truly be the salvation for all of them. However, the fact remains that there are a number of long established businesses in the Britain, the US and elsewhere successfully trading in live worms for various purposes.

The sequential combination of various approaches into treatment trains has become one of the major themes of environmental biotechnology over recent years. AC has particular potential for use in this way and this may prove of increasing importance within the waste management industry in the future. There is a particularly logical fit between this method and traditional composting, since a period of pre-composting permits the thermophilic inactivation of pathogens, while secondary worm action offers a high quality product more rapidly.

There are many advantages to this method, most obviously in that it allows for the input biowaste to undergo established sanitisation procedures without detriment to the worms themselves, which are, as discussed previously, temperature sensitive. A less commonly appreciated benefit of this approach is that the period of initial composting significantly reduces worm ammonia exposure, to which, again, they are very sensitive. However, as with so much of these combined approaches, there is a need to manage the treatment conditions carefully to produce the optimisation desired. There is evidence to suggest that a pre-composting phase has a negative effect on worm growth and reproduction rates (Frederickson *et al.*, 1994) which obviously represents a direct reduction in the overall rate of worm biomass increase. Obviously this has an effect on the overall rate of stabilisation and processing, particularly since it has been demonstrated that the enhanced waste stabilisation achieved under worm treatment is only attained under conditions of high resident worm biomass (Frederickson *et al.*, 1994). It seems reasonable to suggest, then, that to maximise the effectiveness of the combined treatment approach, the initial composting period should be no longer than the minimum necessary to bring about pathogen control of the input biowaste. Though this represents the kind of compromise balancing act so typical of much of environmental biotechnology, it is one which holds much promise. The combination of AC with composting permits both enhanced stabilisation rate and product quality, with the additional bonus that the volatile organic content is also significantly reduced. It is also possible that the natural ability of worms to accumulate various hazardous substances within their bodies will also have implications for waste treatment, particularly if it proves possible to use them deliberately to strip out particular contaminating chemicals.

AC is currently very clearly a minority technology in this role, but the potential remains for it to play a role in the future biological treatment of waste, either as a stand-alone or, as seems more likely, as part of an integrated suite of linked processes. This seems particularly likely if the characteristically superior product derived from this process can be shown to be consistent, since specialist materials in the horticultural and gardening market generally tend to offer better returns. However, only time will tell whether this will prove to be sufficient financial incentive to offset the costs of production and encourage wider adoption of the technology.

Biowaste to ethanol

Around half of the total dry matter in plant origin biomass is cellulose, and since this makes up the majority of the biowaste component in MSW, it represents a huge potential source of renewable energy. As is widely appreciated, sugars can be broken down by certain micro-organisms to produce alcohols, of which ethanol (C_2H_5OH) is the most common. This is, of course a well known application for the production of alcoholic beverages across the world, typically using fermentative yeasts. These organisms are poisoned by ethanol accumulation greater than about 10% which means that, in order to derive higher concentrations, approaches relying on distillation or fractionating are required. In a wider context, ethanol, either as the typical hydrated form (95% ethanol, 5% water) or as azeotropically produced anhydrate, makes a good fuel with excellent general combustion properties.

Historically, the realisation of the huge energy source locked up in the sugars of the cellulose molecule has always been a practical impossibility. The combination of the β 1-4 linkage in cellulose itself, coupled with its typically close association with lignin, making large-scale hydrolysis to sugars a costly and difficult prospect. Some early attempts employed enzymes from wood rotting fungi working on a feedstock of old newspapers or pulp, though the energy involved in actually making the process work often became a limiting factor. In the mid 1990s, various researchers began to investigate the potential of genetically modified bacteria, by inserting appropriate sequences from a variety of naturally occurring wood-rotting organisms. In the following years, a number of technologies have emerged, based both on whole-organism and isolated-enzyme techniques, and the commercial processing of cellulose to alcohol now appears to be about to become a mainstream reality.

A number of countries have begun to show an interest in the potential gains to be had from developing a biowaste-based ethanol industry. Within the US, many individual states have undertaken feasibility studies for their own areas. The California Energy Commission, for example has established that the state-wide annual generation of biowaste exceeds 51 million dry tonnes, comprising forestry residue, MSW and agricultural waste and estimates the resultant maximum potential ethanol yield at more than 3 billion gallons (US).

There are several thriving biomass-to-ethanol production plants elsewhere in the US and the world, though most of these make their alcohol from primary crop plants, rather than biowaste. As with biogas, further discussion of the wider aspects of ethanol and the role of biotechnology in energy production appears in a later chapter.

Eutrophic fermentation (EF)

Eutrophic fermentation (EF) was the name given by one of the authors to the experimental, wet, in-vessel, aerobically enhanced biodegradation process he designed to investigate accelerated decomposition, principally as an alternative to

AD. *Fermentation* in this context relates to the wider application of the word to encompass all microbial breakdown and not simply the anaerobic production of alcohol and *eutrophic* refers to the nutrient-rich environment within which it takes place.

The process arose as a result of research into the enhanced aerated remediation of post AD liquor. The idea of introducing air into liquid or slurrified waste, as was discussed in Chapter 6, is well established as a means of treatment. A laboratory prototype treatment was developed for the specific effluent and was successfully trailed at pilot scale, before subsequently being further developed and extended to the treatment of biowaste.

EF involves rendering the waste into a fine slurry, which is then contained within a bioreactor (Figure 8.4), aerated by means of bubble diffusers at the base of the vessel, mixed and heated to around 35 °C. Within this environment, the input waste breaks down to leave only about 10% of the original volume as recoverable solids within 35 days. The process liquor itself is characteristically less strong than for AD and typically contains only some 6–10% suspended solids. Analysis of this liquid by the Agricultural Development Advisory Service (ADAS) has indicated that it has some potential fertiliser use, based on key performance indicators such as nitrogen, phosphorous and potassium levels, electrical conductivity, generalised nutrient content and heavy metal residues. The results are shown in full in Tables 8.4 and 8.5.

With a 'satisfactory' pH of 6, a 'low' BOD of 2790 mg/l, low ammoniacal nitrogen and the likely slow release of much of the total nitrogen content over several months, ADAS concluded that the liquor was a 'useful source of nitrogen and potash for crop growth'.

Though this method proved itself remarkably efficient in laboratory and intermediate trials, the company for which it was developed subsequently abandoned further research into biological waste treatment and, consequently, EF has never been taken to pilot scale. As a result, it is not possible to say how well it might perform, though for such aerobic processes utilising a completely-mixed,

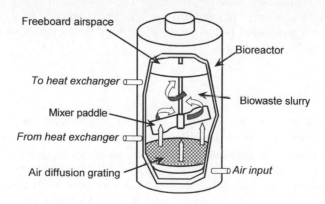

Figure 8.4   *EF bioreactor*

Table 8.4   *EF analysis results (nutrients)*

Principal nutrients	Units per 1000 gal[a]	Comments
Nitrogen (ammoniacal)	2	Low
Nitrogen	5	Moderate
Magnesium	<1	Low
Phosphorus	1	Low
Potassium	6	Moderate

[a]Standard UK agricultural measure for fertiliser value.

Table 8.5   *EF analysis results (metals)*

Metal	mg/l	Comments
Cadmium	<0.25	Low
Chromium	<1	Low
Copper	<1	Low
Lead	<1	Low
Mercury	<0.01	Low
Nickel	<1	Low
Zinc	2.7	Low

suspended growth regime, even relatively small-scale studies tend to be fair representations of full-scale performance. This is particularly relevant to this system, as the hydrolysed organics are directly assimilated by the resident microbes, rather than simply accumulating in the process liquor.

To date, this approach has not made the transition to commercial application, not least as a result of outstanding intellectual property questions. Never-the-less, it stands as a worthwhile example of how alternative technologies can be developed to meet particular needs.

The issue of local circumstance is a recurrent theme throughout the whole of environmental biotechnology and probably so in waste management as much, if not more, than in any other aspect. As a consequence, it is highly unlikely that any one technology will establish itself as the monopoly treatment, which effectively also guarantees the scope for the further development of existing or novel minority biotechnologies.

Closing Remarks

Biowaste management cannot solely be about maximising the diversion of biodegradable material from landfill; it must seek to maximise the return of materials thus diverted back into the chain of utility as an equal priority. This is a particular consideration for those biological treatment processes that result in some kind of compost or similar product. Though it is beyond the scope of this book to discuss the issues of marketing and product quality in depth, it must be obvious that any significant level of biowaste diversion via these approaches demands a sizable and sustainable outlet. Suffice it to say that any such marketing operation requires significant assurances to the customer in terms of the material's quality, safety and value and moves are afoot to establish more widely agreed and accepted criteria. To this end, developments like the application of SOUR testing as a means of objectively assessing microbial activity within the composting matrix may have a wider role to play.

Another area of concern often expressed is that of pathogen persistence, which leads many to view the need for sanitisation as synonymous with sterilisation, which, of course, it is not. The health risks potentially associated with biowaste processing to both workers and end-users has been well documented elsewhere and it is not our intention to restate that work here. However, what is less widely appreciated, and is more directly relevant to our central theme, is that particularly for large scale applications a balanced and thriving community of micro-organisms is one of the most valuable contributions a good biowaste-derived soil amendment can make to poor soils. Removing pathogens, weed seeds and spores from biologically processed waste while not producing a sterile wasteland remains one of the key balancing acts of biowaste treatment.

Ironically, it is the expansion of biotechnology in areas beyond the development of better systems for immediate biowaste treatment which is likely to have major implications for waste management. For any processing technology, there is clear advantage in having a relatively pure input material and this has led to much discussion over the years regarding the respective benefits of separation on site against householder segregation. The latter approach has itself led to the plastic bags in which the biowaste so segregated becoming something of a nuisance at central treatment facilities, typically needing to be opened and removed, which is a labour intensive operation at such a large scale. The increasing use of truly biodegradable plastics has already started to have an impact on the situation, especially at composting plants, since bags which will themselves decompose significantly reduce the amount of work involved.

If the predicted widespread use of cheap bio-plastics, grown as alternative crops in transgenic plants, becomes a reality in the future, then this may itself have repercussions for the amount of material requiring biological treatment. Plastics account for around 8–10% of the developed world's waste stream and, while reducing the demand for finite oil resources for the production of polymers has much to recommend it, one inevitable consequence will be to increase the amount of expressly biodegradable material in refuse. With growing numbers of countries looking to increase biowaste diversion, even allowing for effective recycling initiatives and attempts at waste minimisation, biotechnology may play an even larger part in the approaches to integrated waste management of the future.

The role of such integrated technologies in dealing with waste while simultaneously allowing components, either material or energy, to re-enter the chain of commercial utility will continue to prove a vital one. In whatever form this most fundamental of recycling is achieved, an adequate final market is essential. Whether the end product is fundamentally reclaimed humus and minerals or a usable fuel, without a genuine and environmentally beneficial end-use of some kind, the whole operation becomes little more than an exercise in shifting the problem from one place, to another.

References

Bess, V. (1999) Evaluating the Microbiology of Compost. BioCycle Magazine (May 1999), pp. 62–63.

Cope, C. (1995) An Introduction to the Chemistry of Landfill Gas and Leachate, Centre for Environmental Research and Consultancy (CERC), Cambridge.

Denham, C. (1996) Large Scale Profitable Manure Disposal, Booklet 2 Wonder Worms, Sowerby Bridge.

DETR (Department of the Environment, Transport and the Regions) (1999a) Limiting Landfill: A Consultation Paper on Limiting Landfill to Meet the EC Landfill Directive's Targets for the Landfill of Biodegradable Municipal Waste, Crown copyright, pp. 11–12.

DETR (Department of the Environment, Transport and the Regions) (1999b) A Way with Waste; A Draft Strategy for England and Wales, Part 2, Crown copyright, p. 90.

Dolk, H., Vrijheid, M., Armstrong, B. *et al.* (1998) Risk of congenital anomalies near hazardous-waste landfill sites in Europe: the EUROHAZCON study. *Lancet*, **352**, 423–427.

Elliott, P., Briggs, D., Morris, S. *et al.* (2001) Risk of adverse birth outcomes in populations living near landfill sites. *British Medical Journal*, **323**, 363–368.

Frederickson, J., Butt, K., Morris, R. and Daniel, C. (1994) Combining vermiculture with traditional green waste composting systems. Presented at the 5th International Symposium on Earthworm Ecology, July 5–9, Columbus, OH.

Frostell, B. (1992) The role of biological waste treatment in integrated waste management. Proceedings of Biowaste '92, ISWA/DAKOFA, Copenhagen, p. 4.

Lemmes, B. (1998) The 'Tao' of organics. *Wastes Management*, 18.

Lynch, V., Riley, M., Davey, M.A. and Thursfield, V. (2006) Investigating Health Concerns in Populations Living Near the Cleanaway Landfill in Tullamarine. Environmental Health Unit, Public Health Branch, Victorian Government Department of Human Services, Melbourne.

Menon, S. (1994) Worms recruited to clear Bombay's rubbish. *New Scientist*, 5.

Monserrate, E., Leschine, S.B. and Canale-Parola, E. (2001) *Clostridium hungatei* sp. nov., a mesophilic, N_2-fixing cellulolytic bacterium isolated from soil. *International Journal of Systematic and Evolutionary Microbiology*, **51**, 123–132.

Vrijheid, M. (2000) Health effects of residence near hazardous waste landfill sites: a review of epidemiologic literature. *Environmental Health Perspectives*, **108**, 101–112.

Vrijheid, M., Dolk, H., Armstrong, B. *et al.* (2002) Chromosomal congenital anomalies and residence near hazardous waste landfill sites. *Lancet*, **359**, 320–322.

Wright, M. (1998) Home composting: real waste minimisation or just feel good factor? *Wastes Management*, 27–28.

Case Study 8.1   Composting for Methane Avoidance (Telipok, Malaysia)

Composting the putrescible component of MSW has found favour in many countries as a means to reduce both the amount of material entering landfills and simultaneously deal with the problem of methane production. The Kota Kinabalu Composting Project – the first of its kind in Malaysia – provides an excellent example of this approach.

The operation occupies about 7 ha (18 acres) of land, lying entirely within the boundaries of the operational Kayu Madang sanitary landfill, near the town of Telipok some 30 km to the northeast of the city of Kota Kinabalu and 2 km inland from the South China Sea. Receiving an up-front 500 tonnes/day of residential and commercial MSW from Kota Kinabalu and the neighbouring districts of Kota Belud, Tuaran and Penampang, on-site separation within the purpose-built plant subsequently yields around 300 tonnes/day of biodegradable waste for composting. The mixed waste is delivered to the plant by collection vehicles and bulk items are then removed from the processing stream by hand. The material is then transferred by front-end loaders to the new sorting facility, where the recyclable components, principally plastics, glass, metals and paper, are extracted by a combination of mechanical systems and manual picking to produce a default organic feedstock which is then arranged in windrows. After composting, the product is screened, any remaining inerts being removed for disposal at the adjacent landfill.

SMART Recycling, the plant's operators, anticipate the production of 45 000 tonnes/year of compost for use as a soil conditioner, along with an estimated greenhouse gas reduction of around 740 000 tonnes of CO_2 equivalent over its 10-year lifespan from 2009 to 2019. This aspect is of particular significance to the economics of the project. Capital expenditure and running costs are always a factor in composting operations, typically depending on the revenue generated by gate fees and product sales to make them commercially viable. Implementing the scheme in a country which has no existing established market for the compost and where no gate fees are to be paid would have been impossible without the emission reduction purchase agreement between the World Bank's Carbon Fund for Europe and the operators for the purchase of 340 000 tonnes of CO_2-equivalent of Certified Emissions Reductions up to the end of 2014.

Case Study 8.2   Mixed Anaerobic Digestion (Mantova, Italy)

The potential that AD has for recovering a substantial energy yield from material with a high organic content inevitably makes it an attractive option for dealing with waste. As Rivalta Energy's plant outside of Mantova demonstrates, a carefully designed AD facility can integrate very successfully with a variety of existing environmental strategies and waste management arrangements.

With a design capacity of 10 000 tonnes/year of municipal/commercial and agricultural feedstocks, the plant principally comprises three individual $70\,m^3$ pre-treatment compartments, a complete mix digester and four $350\,kW$ generators, currently producing $1\,MW$ of electricity, but built to be expandable to $2\,MW$. Municipal and commercial solid wastes are processed into a slurry within their receiving compartment, with the addition of digestate or dairy manure to aid the mixing, and a sophisticated biofilter system is used for odour control.

Producing a suitably homogeneous mix for digestion takes between 15 and 60 minutes, depending on the nature of the incoming waste, after which the slurry, now at around 10–15% solid content, is pumped into a cyclone separator, where any remaining entrapped non-organic inclusions are removed to feed into existing recycling arrangements as appropriate. The slurry then flows to the third compartment, where it is balanced and blended with agricultural manures and other input feedstocks before being pumped into the digester. The Rivalta digester is a completely mixed system, consisting of two primary and two secondary cells. Biogas mixers within the cells ensure that the digesting substrates remain completely homogenised throughout the process to optimise performance in terms of both biological action and biogas yield.

Once the digestion process has been completed the material is dewatered using sequential screw press and roller press separation, with the solids being sent for final aerobic composting, while the liquid is held in a storage tank and subsequently used either to wet incoming solid waste in the pre-treatment phase, or applied to local fields. After composting and maturation, the solid material is also used for agriculture.

The success of this facility, which was completed and commissioned in 2007, has led Rivalta Energy to developed three further plants using the same technology.

9

Genetic Manipulation

Genes have been manipulated by man for a very long time, that is, if selective breeding, which has been practised for centuries in agriculture and elsewhere to develop desirable characteristics in domesticated animals and plants, is to be considered as manipulation, as it rightly should. Even from the early days of Gregor Mendel, the Moravian monk and pioneer of genetic analysis, plants were bred to bring out interesting, useful and sometimes unusual traits. Many of these are now lost to classical plant breeders because of divergence of strains leading to infertile hybrids. One of the joys of genetic engineering (GE) is that in some cases, ancient genes may be rescued from seed found in archaeological digs, for example and reintroduced by transfer into modern strains. It has been proposed that the exchange of genetic information between organisms in nature is considerably more commonplace than is generally imagined (Reanney, 1976; Hehemann *et al.*, 2010) and could explain the observed rates of evolution. In bacteria, the most likely candidates for genetic transfer are plasmids and bacteriophage, and since eukaryotes lack plasmids, their most plausible vectors are eukaryotic viruses. This, of course, is in addition to DNA transfer during sexual reproduction. Existing knowledge would suggest that exchange involving a vector requires compatibility between the organism donating the genetic material, the vector involved and the recipient organism. For example, two bacteria must be able to mate for plasmid transfer to take place, or if a virus is involved as a vector, it must be able to infect both the donor and recipient cells or organisms. However, there is evidence to suggest that this view is somewhat naive and that there is considerably more opportunity for genetic exchange between all cells, prokaryotic and eukaryotic, than is popularly recognised. This idea, proposed by Reanney (1976) is developed in Chapter 3.

Bacteria are notorious for their ability to transfer genes between each other as the need arises thanks to the location on plasmids of most of the gene groups, or operons, involved in breakdown of organic molecules. Strong evidence for the enormous extent of these 'genomic pools' comes from analysis of marine sediment (Cook *et al.*, 2001). Throughout this book, the point has been made that microorganisms involved in remediation do so in their 'natural' state largely because they are indigenous at the site of the contamination and have developed suitable capabilities without any external interference. However, sometimes after

Environmental Biotechnology: Theory and Application, Second Edition Gareth M. Evans and Judith C. Furlong
© 2011 John Wiley & Sons, Ltd.

a sudden contamination such as a spill, microbes are not able to amass useful mutations to their DNA quickly enough to evolve suitable pathways to improve their fitness for that changed environment, and so they may be 'trained' by the artificially accelerated expansion of pre-existing pathways. The final option is that they may be genetically engineered. Organisms which represent the 'norm', frequently being the most abundant members occurring in nature, are described as 'wild type'. Those with DNA which differs from this are described as mutant. Alteration can be by the normal processes of evolution which constantly produce mutants, a process which may be accelerated artificially, or by deliberate reconstruction of the genome, often by the introduction of a gene novel to that organism. This latter route is the basis of GE which has several advantages over traditional breeding or selection techniques. The process is specific, in that one gene, or a selected group of genes, is transferred and so the mutation is quite precise. There is flexibility in the system in that, depending on the modifications made to the genome, a new product may be produced or the level of expression of the existing product or products may be altered in quantity or proportions to each other. This whole subject of proteomics is a discipline in its own right on which volumes are published. Another advantage often quoted is that GE allows genes to be transferred between totally unrelated organisms. The preceding discussion suggests that this is not a phenomenon unique to GE, but it is at least defined and specific.

Training: The Manipulation of Bacteria Without Genetic Engineering

A general procedure is to take a sample of bacteria from, at, or near, the site of contamination from which a pure culture is obtained in the laboratory and identified, using standard microbiology techniques. The 'training' may be required either to improve the bacterium's tolerance to the pollutant or to increase the capabilities of pathways already existing in the bacterium to include the ability to degrade the pollutant, or a combination of both. Tolerance may be increased by culturing in growth medium containing increasing concentrations of the pollutant so that, over successive generations, the microbe becomes more able to withstand the toxic effects of the contaminant. Reintroduction of these bacteria to the polluted site should give them an advantage over the indigenous bacteria as they would be better suited to survive and remediate the contamination. Improving the microbes ability to degrade a contaminant, sometimes referred to as catabolic expansion, may be increased by culturing the bacteria in growth medium in which the contaminant supplies an essential part of the nutrition, such as being the only carbon source. Only bacteria which have undergone a mutation enabling them to utilise this food source will be able to survive and so the method effectively selects for the desired microbe; everything else having died.

It has been argued that under laboratory conditions where cultures of bacteria are isolated from each other to prevent cross-contamination, mutations are most likely to occur as a result of an error in DNA replication. This is far less likely to be the most prominent source of mutation in nature, as the microbes are constantly in close proximity with other organisms and, consequently, the opportunity for exchange of genetic material is enormous. In fact the process of DNA replication has a very high fidelity, the reasons for which are obvious. An increased rate of error may be forced upon the organism, speeding up the rate of mutation, by including a mutagen in the growth medium. A mutagen is a chemical which increases the rate of error in DNA replication, often by causing a very limited amount of damage to the DNA such that the DNA polymerase, the enzyme responsible for synthesising DNA, is unable to determine the correct base to add into the growing nucleotide chain. If the error in the naescent strand cannot be recognised and corrected, the fault becomes permanent and is handed on through the generations.

Manipulation of Bacteria by Genetic Engineering

Genetic manipulation by the deliberate introduction of defined genes into a specified organism is a very powerful, established technique in constant development, some times at phenomenal rates of progress. The techniques have produced some exciting hybrids in all areas of research, both microscopic; bacteria and fungi, usually described as recombinants, and macroscopic; principally higher plants and animals, commonly described as transgenics. The latter term refers to the principal of deliberate transfer of a gene from one organism to another in which it is not normally resident. This earns the incoming gene the title of 'foreign'. Some examples of these which are relevant to environmental biotechnology will be discussed later in this chapter.

   Some of the developments are of great potential interest and represent some exciting and innovative work. However, it must be said that, in practice, a very tiny proportion of all endeavour in the name of environmental biotechnology has, or is likely to have in the future, a direct reliance for its effectiveness on the type of recombinants and transgenics currently being developed. This is not because of the limits of GE, which in principal are almost boundless, given sufficient resources, but because of cost. It is a principal factor as the technology and research to produce transgenic organisms attracts an inherently high price. While such a situation may be sustainable by pharmaceutical companies and perhaps to a lesser extent, agribiotechnology companies possibly able to command a high return on sales of the product, it is rarely sustainable in applications of environmental technology. Few commercial organisations are excited at the prospect of spending a large proportion of their income on waste disposal, for example and will normally only do so when absolutely necessary.

   There are other factors which affect the suitability of transgenic organisms in this science due to current requirements for containment. In addition, the way in which such a recombinant is utilised may cause problems of its own. For example, if the recombinant is a microorganism structured to improve the rate of degradation of a pollutant, its performance may be exemplary in laboratory conditions but when it is applied in bio augmentation it is in competition with indigenous species which could outgrow the recombinant. The novel bacterium may also lose its carefully engineered new capability through normal transfer of genes given the high level of promiscuity between bacteria. A highly controlled and contained environment such as a bioreactor may circumvent some of these objections but it is not always practical to move the contamination to the solution rather than the solution to the contamination. Again this involves expense and practical considerations, not least of which are safety concerns associated with the transport of contaminated material.

In reality, there is rarely any need to use recombinants or transgenics and it is far more likely that the required metabolic capability will be provided by indigenous organisms, or ones which have been trained for the task. There are, however, some exotic and ingenious applications, and by way of illustration, some examples are given here. The aim is to provide an overview of some of the more frequently used technologies together with specific examples. There are very many excellent textbooks and specialised publications which should be consulted should a more detailed and working knowledge be required. However, an overview of the principles of GE are given here for the benefit of those unfamiliar with the technology.

Basic Principles of Genetic Engineering

There are endless permutations of the basic cloning procedures but they all share some fundamental requirements. These are: the enzymes, solutions and equipment necessary to perform the procedures; the desired piece of DNA to be transferred; a cloning vector; and the recipient cell which may be a whole organism. For the process to be of any measurable value, it is also essential to have some means of determining whether or not the transfer has been successful. This is achieved by the use of marker genes. The requirements referred to above are described in the following sections.

Enzymes, solutions and equipment

There are many steps involved in the isolation of DNA which now have become standard laboratory techniques. Once DNA has been isolated from an organism, it is purified from contaminating material such as protein and is precipitated out of aqueous solution by the addition of alcohol, for example ethanol, to approximately 70%. The DNA appears as a white, semi-transparent material, coiling out of solution on addition of the alcohol. This may be collected by centrifugation and dried down ready for the next stage which is usually enzyme digestion. The aim of the next stage is to insert the DNA into the vector, for which the ends of the DNA and the vector have to be prepared. This may be done by restriction endonucleases which recognise specific sequences within the DNA and cut at

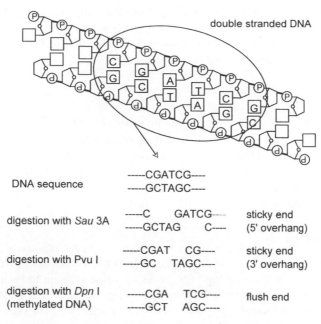

double stranded DNA		

DNA sequence	-----CGATCG---- -----GCTAGC----	
digestion with *Sau* 3A	-----C GATCG---- -----GCTAG C----	sticky end (5' overhang)
digestion with Pvu I	-----CGAT CG---- -----GC TAGC----	sticky end (3' overhang)
digestion with *Dpn* I (methylated DNA)	-----CGA TCG---- -----GCT AGC----	flush end

Figure 9.1   *Restriction enzymes*

that site, either producing a flush or staggered end, Figure 9.1, or by incubation over a very limited time period with an exonuclease which digests the end of the DNA and followed by further digestion with another nuclease which tidies up the ends to produce flush ends. There are other restriction nucleases which recognise a site in DNA but cut at some distance from it, but these are rarely of any value in cloning procedures.

Preparation of the vector is dictated by the type of end prepared for the insert DNA; flush or 'sticky'. If it is flush, it does not much matter how that was achieved so long as the vector receiving it is also flush, but if it is sticky, the appropriate sticky end must be prepared on the vector by a suitable restriction endonuclease. There are many methods of DNA and vector preparation, all of which have their advantages and disadvantages and, although interesting in themselves, are beyond the scope of this book.

Having prepared the ends, the next step is to stick the pieces together. The prepared insert, or 'foreign' DNA is incubated with the prepared vector in an aqueous solution containing various salts required by the enzymes, and ligase which is an enzyme the function of which is to make the bond between the free phosphate on a nucleotide base and the neighbouring ribose sugar thus 'repairing' the DNA to make a complete covalently linked chain. This recombinant DNA molecule may be transferred into a cell where it undergoes replication in the usual way. If the DNA is not viral, introduction will be by direct entry through the cell membrane achieved by any one of a number of standard techniques all based on making the membrane permeable to the DNA molecule. However, if the 'foreign' DNA is part of a recombinant virus, it has to be packaged into particles, and then transferred into cells by infection. A check may be made on the product by carrying out analyses described later.

DNA for transfer

Most commonly, this is a piece of double stranded DNA which contains the coding sequence for a gene. It may have been obtained from a number of sources, for example genomic DNA, a complementary DNA (cDNA) library, a product of a polymerase chain reaction (PCR) or a piece of DNA chemically produced on a DNA synthesiser machine. Another source is from a DNA copy of an RNA virus as in the replicative form of RNA viruses.

Genomic libraries

Genomic DNA, in this context, is material which has been isolated directly from an organism, purified and cut up into pieces of a size suitable to be inserted into a cloning vector. These pieces may either be ligated as a total mixture, into a suitable vector to produce a genomic library, or a specific piece may be isolated and prepared as described above. Genomic libraries are very useful, as they may be amplified, and accessed almost limitlessly, to look for a specific DNA sequence thus reducing the amount of work involved in any one experiment. The disadvantage is that if the genomic library is of a eukaryotic origin, which is almost exclusively the case, the genes will contain regions, or introns, which are quite normally inserted along its length and are processed out of the RNA copy during maturation prior to protein synthesis. This is a problem if the gene is to be expressed, in other words, if the protein is to be made from the DNA blueprint. Prokaryotics do not contain introns in their genes and so do not possess the mechanisms for their removal. Furthermore, introns are not necessarily processed correctly even if the expression system is eukaryotic. This problem can be avoided by using a cDNA instead of a genomic library.

cDNA libraries

In eukaryotes, the first product of transcription from DNA is not messenger RNA (mRNA) but heterogeneous nuclear RNA (hnRNA). This is mRNA prior to the removal of all the non-coding sections, or introns, which are discarded during the processing to produce the mature mRNA. cDNA is DNA which has been artificially made using the mature mRNA as a template, which is then used as the template for the second strand. Thus the synthetic DNA product is simply a DNA version of the mRNA and so should overcome the problems of expression outlined above.

Polymerase chain reaction

The PCR is a powerful technique which amplifies a piece of DNA of which only a very few copies are available. The piece must be flanked by DNA whose sequence is known or at least a close approximation can be guessed. This allows a short sequence of DNA to be synthesised of only a few nucleotides long, to bind specifically to the end of the sequence and act as a primer for the DNA polymerase to make one copy of the whole piece of DNA. A second probe is used for the other end to allow the second strand to be synthesised. The process is repeated by a constant cycling of denaturation of double stranded DNA at elevated temperature to approximately 95 °C, followed by cooling to approximately 60 °C to allow annealing of the probe and complementary strand synthesis. This technique requires the use of DNA polymerases able to withstand such treatment. Two bacteria from which polymerases have been isolated for this purpose arc *Thermococcus litoralis* and *Thermus aquaticus*. This latter extremophile has been discussed in Chapter 3.

Cloning vectors

A cloning vector is frequently a plasmid or a bacteriophage (bacterial virus) which must be fairly small and fully sequenced, able to replicate itself when reintroduced into a host cell, thus producing large amounts of the recombinant DNA for further manipulation. Also it must carry on it 'selector marker' genes. These are different from the reporter genes described below which are indicators of genomic integrity and activity. A common design of a cloning vector is one which carries two genes coding for antibiotic resistance. The 'foreign' gene is inserted within one of the genes so that it is no longer functional therefore it is possible to discriminate by standard microbiology techniques, which bacteria are carrying plasmids containing recombinant DNA and which are not. Selector genes may operate on at least two levels, the first at the level of the bacterium, usually *Escherichia coli*, in which the manipulations are being performed described above and the second being at the level of the final product, for example a higher plant. In this case such a selector gene can be one to confer resistance to antibiotics like kanamycin or hygromycin.

Standard cloning vectors normally carry only selector marker genes required for plasmid construction. To make the manipulations easier, these genes normally contain a multi cloning site (MCS) which is a cluster of sites for restriction enzymes constructed in such a way to preserve the function of the gene. Disruption by cloning into any one of these sites will lose the function of that gene and hence, for example if it codes for antibiotic resistance, will no longer protect the bacterium from that antibiotic. An example is shown in Figure 9.2. This is pGEM® (Promega, 1996) which has a MCS in the β-*gal* gene. This codes for β-*galactosidase* from the *E. coli lac* operon, which has the capacity to hydrolyse

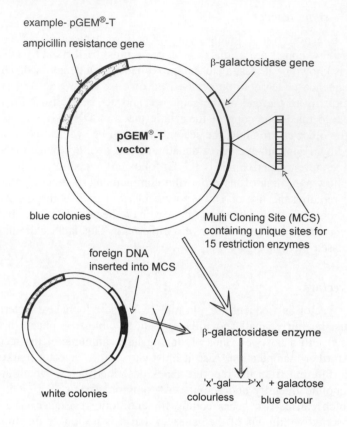

example- pGEM®-T

ampicillin resistance gene

β-galactosidase gene

pGEM®-T
vector

Multi Cloning Site (MCS)
containing unique sites for
15 restriction enzymes

blue colonies

foreign DNA
inserted into MCS

β-galactosidase enzyme

white colonies

'x'-gal ⟶ 'x' + galactose
colourless   blue colour

pGEM®-T is a registered tradename (Promega)

Figure 9.2   *Cloning vector*

x-gal, a colourless liquid, to produce free galactose and 'x' which results in a blue pigment to the colony. Thus the screening for successful insertion into the MCS is a simple scoring of blue (negative) or white (possibly positive) colonies. The success of the experiment can be determined quickly as this cloning vector also has sequences at either side of the MCS which allows for rapid DNA sequencing.

Additionally, some eukaryotic viruses may be used as vectors but these tend to be so large that direct cloning into them is difficult. A solution to this is to carry out manipulations on the desired DNA fragment cloned into a bacterial plasmid and then transfer the engineered piece into the virus thus making a recombinant eukaryotic virus. One such virus now used extensively in GE, both as a cloning vehicle and as an excellent expression vector, is Baculovirus illustrated in Figure 9.3 which is also effective as a bioinsecticide.

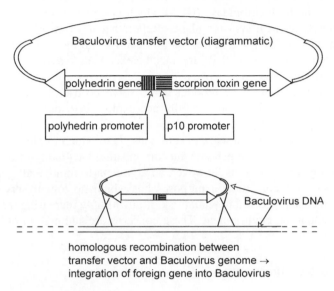

example-
introduction of scorpion toxin for an improved
Baculovirus insecticide. Stewart et al (1991)

Figure 9.3 *Recombinant Baculovirus*

Expression vectors

These are similar to the vectors described above but in addition have the required signals located before and after the 'foreign' gene which direct the host cell to translate the product of transcription into a protein. It is sometimes a difficult, expensive or time consuming procedure to analyse for product from the 'foreign' gene and so, in addition to the selector genes described above, there are frequently reporter genes to indicate whether or not the signals are 'switched on' allowing the 'foreign' DNA to be expressed. There are many reasons which are difficult to predict, why even a perfectly constructed gene may not be functional, such as the consequence of the exact site of insertion in the genome; hence the need for inbuilt controls.

Reporter genes

There are many such genes in common use and these usually code for an enzyme. The most common is β-*galactosidase*, mentioned above. This enzyme, supplied with the appropriate reagents, may also catalyse a colour change by its activity on a variety of chemical compounds typified by orthonitrophenolgalactoside (ONPG) which changes from colourless to yellow on hydrolysis in much the same way as the blue/white screening described above for the cloning vector, pGEM®. Other reporter genes produce enzymes which can cause the emission of light such as the *luciferase* isolated from fireflies, or whose activity is easy and quick to assay like the bacterial β-glucuronidase (GUS) which is probably the most frequently used reporter genes in transgenic plants. Reporter genes can only be a guide to the process of transcription and translation occurring in the cell and it has been acknowledged for some time that care must be exercised to avoid misinterpreting data (Pessi, Blumer and Haas, 2002).

As with selector genes, the reporter genes serve no useful purpose once the cloning procedure has been successfully accomplished to produce the finished product. In the early days of this technology, these genes would normally be left *in situ* to avoid the extra work of removing them which might also upset the structure of the recombinant genome thus diminishing the quality of the carefully engineered organism. There is, however, an argument to remove all genes which were necessary for construction purposes but which no longer serve a useful purpose, to reduce perceived potential risk of unwittingly increasing the spread of genes throughout the environment. These concerns are addressed in Chapter 10.

Analysis of Recombinants

The design of the plasmid was such that insertion of 'foreign' DNA allows for a colour test, or causes a change in antibiotic sensitivity, either to resistance (positive selection) or sensitivity (negative selection). This constitutes the first step in screening. The second stage is usually to probe for the desired gene using molecules which will recognise it and to which is attached some sort of tag, usually radioactive or one able to produce a colour change. The next stage is normally to analyse the DNA isolated from possible recombinants, firstly by checking the size of the molecule or pieces thereof, or by sequencing the DNA. This is the most informative approach but used to be very laborious. With the current and ever developing automated protocols, DNA sequencing has become a standard part of recombinant analysis procedure. However, if a large number of samples are to be analysed it is usually quicker and cheaper to scan them by a procedure described as a Southern blot, after Ed Southern, the scientist who designed the technique.

In this procedure, the DNA is spread out by electrophoresis on a gel which is then probed by a piece of DNA complementary to the sequence of interest. If a band shows up on autoradiography then the probe has found a mate and the required sequences are present, at least in part. DNA sequencing is then required to confirm exactly what has occurred during the cloning procedure but the advantage is that only the samples which are very likely to contain the required insert are sequenced, thus saving time and expense.

From this technique, the Northern blot has developed, which is much the same idea except that the material spread out on the gel is RNA rather than DNA, and the Western blot which is slightly different in that the material electrophoresed is protein, which is probed with antibodies against the anticipated protein, rather than nucleic acid, as is the case in Southern and Northern blots.

Recombinant Bacteria

GE of microorganisms for use in environmental biotechnology has tended to focus on the expansion of metabolic pathways either to modify the existent metabolic capability or to introduce new pathways. This has various applications, from the improved degradation of contaminants, to the production of enzymes for industry, thus making the process less damaging to the environment. One such experimental example taken from 'clean technology' with potential for the manufacturing industry, is a strain of *Escherichia coli* into which was engineered some 15 genes originating from *Pseudomonas*. These were introduced to construct a pathway able to produce indigo for the dyeing of denim, commented on by Bialy (1997). The traditional method requires the use of toxic chemicals with the associated safety measures and inherent pollution problems. Similar technologies were investigated in the early 1980s, by Amgen in the USA and Zeneca in the UK, but were not pursued due to questionable profitability. Whether or not this route will ultimately be taken up by industry still remains to be seen (BMB, 1995).

Recombinant Yeast

Yeast, being unicellular eukaryotes, has become popular for cloning and expressing eukaryotic genes. It is fairly simple to propagate, some species being amenable to culture in much the same way as bacteria. Yeast cells are surrounded by a thick cell wall which must be removed to permit entry of DNA into the cell. There are several types of plasmid vector available for GE, some of which have been constructed to allow replication in both bacteria and yeast (Beggs, 1981). All have a region which permits integration into the host yeast genome by recombination. This occurs by alignment of the sequences complementary between the host genome and the incoming plasmid DNA. Two crossover events then take place which effectively swap over a piece of host DNA with the plasmid DNA. A similar process occurs in the construction of recombinant Baculoviruses.

Recombinant Viruses

The insect virus, Baculovirus, has been shown to be the method of choice for the over expression of genes in many applications of molecular biology. The viral genome is large relative to bacterial plasmids and so DNA manipulations are normally carried out on a plasmid maintained in *Escherichia coli*. Introduction of the reconstructed gene, or group of genes, to the Baculovirus DNA occurs by recombination in much the same way as described for the formation of recombinant yeast. One example of interest to environmental biotechnology is the replacement of p10, one of the two major Baculovirus proteins, polyhedrin being the other, by the gene for a scorpion neurotoxin, with a view to improving the insecticidal qualities of the virus, sketched in Figure 9.3 (Stewart *et al.*, 1991). The 'promoters' at the start of the gene, referred to in the figure, are the regions of RNA which regulate protein synthesis, from none at all, to maximum expression.

Transgenic Plants

Currently, GE in agribiotechnology is focusing on genetic modifications to improve crop plants with respect to quality, nutritional value and resistance to damage by pests and diseases. Other avenues under investigation aim to increase tolerance to extreme environmental conditions, to make plants better suited for their role in pollutant assimilation, degradation or dispersion by phytoremediation, or to modify plants to produce materials which lead to the reduction of environmental pollution. Crop quality improvements such as the control of fruit ripening (Grierson and Schuch, 1993), an example of which is the oft quoted, Flavr-Savr tomato, and the production of cereals with improved nutritional value, are not addressed here because, although of great interest to the food industry, they are of more peripheral relevance to environmental biotechnology. Many of the transgenic plants, examples of which are given later in this chapter, have been produced using the Ti plasmid transfer system of *Agrobacterium tumefaciens* and are often used together with the 35S CaMV promoter. Both of these tools are described from a GE techniques viewpoint in this chapter and from a biological standpoint in Chapter 10.

Transformation of plants

There are two practical problems associated with GE of plants which make them more difficult to manipulate than bacteria. Firstly they have rigid cell walls and secondly they lack the plasmids which simplify so much of GE in prokaryotes. The first problem is overcome by the use of specialised techniques for transformation, and the second by performing all the manipulations in bacteria and then transferring the final product into the plant. The DNA construct contains regions of DNA which are complementary to the plant DNA to enable the inserted piece to recombine into the plant genome.

The most popular method of transforming plants is by the Ti plasmid but there are at least two other methods also in use. The first is a direct method where DNA is affixed to microscopic bullets which are fired directly into plant tissue. An example of this technology is the introduction into sugarcane of genes able to inactivate toxins produced by the bacterium, *Xanthomonas albilineans*, causing leaf scald disease (Zhang, Xu and Birch, 1999). This method of biolistic bombardment, may increase in popularity in line with improvements to plastid transformation. It is now possible to produce fertile transgenics expressing foreign proteins in their edible fruit (Ruf *et al.*, 2001).

The second is by protoplast fusion which is a process whereby the plant cell wall is removed leaving the cell surrounded only by the much more fragile membrane. This is made permeable to small fragments of DNA and then the cells allowed to recover and grow into plants. These methods can be unsuccessful due to difficulties in recovery of the cells from the rather traumatic treatments and also because the DNA introduced has a tendency to be inserted randomly into the genome, rather than at a defined site. However, both methods enjoy the advantage that DNA enters the cell exactly as constructed and has not passed through an intermediate vector giving the opportunity for gene rearrangement.

Transformation by the Ti plasmid of *Agrobacterium tumefaciens*, shown diagrammatically in Figure 9.4, suffers from few disadvantages other than the limitation that it does not readily infect some cereal crops. This potential problem

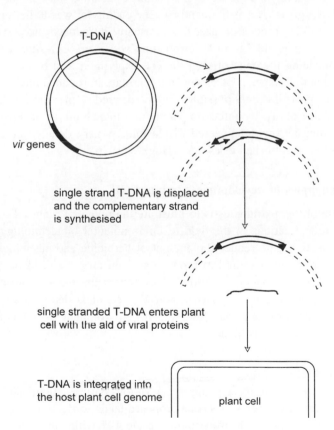

Figure 9.4 *Ti plasmid of* Agrobacterium tumefaciens

has been addressed by attempting to increase its host range (Godwin, Fordlloyd and Newbury, 1992) which has met with success, leading to improved transformation procedures (Le *et al.*, 2001). In essence, the wild type plasmid contains genes which causes the transfer of a piece of DNA, 'T-DNA', into a plant cell. This piece is bordered by sequences of 24 base pairs in length which are repeats of each other. This structure is fairly common in DNA and is described as a direct repeat. The T-DNA comprises genes which cause crown gall disease. These genes may be cut out and replaced by DNA containing the gene of choice to be introduced into the plant. There are many additional elements which may be included in the construct. For example, if the aim is to express the gene, it is preceded by a strong promoter, most commonly the '35S' promoter of Cauliflower Mosaic Virus.

In addition to the above, it is important to know if the 'foreign' gene is being expressed and so frequently a 'reporter' gene described in the section above is also included located close to the gene of interest. Recombination is not 100% efficient, and so a method of selection is required such that only plants containing the novel DNA grow. This is frequently a gene coding for weedkiller or antibiotic resistance. On the grounds of size, this is usually introduced more successfully on a second Ti plasmid during a co-infection with *Agrobacterium* carrying the plasmid containing the gene of interest. The experiment can become somewhat complicated at this stage, as other selector genes are introduced into the plasmids to ensure that growth is only possible if all the desired elements are present in the plant cell. This can involve infection with two or three cultures of *Agrobacterium* each containing its own engineered Ti plasmid. A very detailed description of the Ti plasmid is published elsewhere (Hughes, 1996).

Selected examples of developments in plant GE

The purpose of these examples is to illustrate the potential plant GE could hold for practical applications in the field of environmental biotechnology. In some cases the intention is to reduce the amount of herbicide and pesticides, or other agricultural chemicals required to produce a given crop yield, in others it is to improve tolerance of harsh conditions or to protect the plants from attack thus reducing wastage. The intention is to note the technical details here, while the effects such developments may have on the environment as a whole, feature elsewhere throughout this book.

Broad range protection

A general strategy to protect plants from various viruses, fungi and oxidative damage by a range of agents, has been proposed using tobacco plants as a model. The transgenics express the iron-binding protein, ferritin, in their cells which appears to afford them far ranging protection (Deák *et al.*, 1999).

Resistance to herbicides

'Glyphosate', one of the most widely used herbicides, is an analogue of phos-phoenol pyruvate and shows herbicidal activity because it inhibits the enzyme *5-enolpyruvylshikimate-3-phosphate synthase*. The gene coding for this enzyme has been identified, isolated and inserted into a number of plants including petunias. In this case, the gene was expressed behind a CaMV promoter and intro-duced using *A. tumefaciens*, leading to very high levels of enzyme expression. As a consequence, the recombinant plants showed significant resistance to the effects of glyphosate (Shah *et al.*, 1986). Developments in this strategy include the formation of a chimaeric *synthase* enzyme, the analysis of which should lead to improved herbicide resistance in transgenic crops using this strategy (He *et al.*, 2001).

An alternative approach but still using *A. tumefaciens*, has been to transfer the genes for mammalian cytochrome P450 monooxygenases, known to be involved in the detoxification (and activation) of many xenobiotics including pesticides, into tobacco plants. These transgenics displayed resistance to two herbicides, *chlortoluron* and *chlorsulphuron* (Yordanova, Gorinova and Atanassov, 2001).

Improved resistance to pests

Plants have an inbuilt defence mechanism protecting them from attack by insects but the damage caused by the pests may still be sufficient to reduce the com-mercial potential of the crop. The usual procedure is to spray the crop with insecticides but in an effort to reduce the amount of chemical insecticides being used, plants are being engineered to have an increased self-defence against pests. Attack by insects not only causes damage to the plant but also provides a route for bacterial or fungal infection in addition to the role played in the spread of plant viruses. With a view to increasing resistance to sustained attack, the genes coding for the δ-endotoxin of the bacterium, *Bacillus thuringiensis*, described a little more fully in Chapter 10, have been transferred into plants. Examples are of a synthetic *B. thuringiensis* δ-*endotoxin* gene transferred, in the first case, by *A. tumefaciens* into Chinese cabbage (Cho *et al.*, 2001) and in the second, by biolistic bombardment into maize (Koziel *et al.*, 1993). In both cases, the trans-genic plants showed greatly improved resistance to pest infestation. There are, however, some problems with crop performance of some genetically engineered plants highlighted in Magg *et al.* (2001). Insects are able to develop resistance to Bt products which is a problem addressed by insertion of δ-*endotoxin* genes into the chloroplast genome rather than into that of the plant's nucleus, with promising results (Kota *et al.*, 1999; Daniell and Moar, 2007).

It may be recalled that for each amino acid incorporated into a protein there is usually a choice of three or four codons all of which code for that same amino acid. Different organisms have distinct preferences for a particular codon, thus *Bacillus thuringiensis* tends to use codons richer in thymidine and adenine than the plant cells into which the gene is placed. There are also signals controlling the expression of these genes relevant to bacteria, rather than eukaryotes, which will not function very well, if at all, in the plant cell. For these reasons, expression may benefit from modification of the DNA sequence to compensate for these differences while maintaining the information and instructions. This may account in part for the very high levels of expression and stability of the Bt proteins whose genes have been introduced, by (biolistic) microbombardment, into chloroplasts (De Cosa *et al.*, 2001) which, because of their prokaryotic ancestry, have 'protein synthesising machinery' more in keeping with prokaryotes than the eukaryotic cell in which they co-habit.

Attempts to improve virus resistance have lead to the introduction, by *A. tumefaciens*, of the genes expressing antibodies to the coat protein of Tobacco Mosaic Virus (TMV). Expression of these in the plant lead to complete immunity against TMV (Bajrovic *et al.*, 2001).

Improved resistance to disease

Bacteria communicate with each other by way of small diffusible molecules such as the N-acylhomoserine lactones (AHLs) of Gram negative organisms. In this way, described as 'quorum sensing', they are able to detect when a critical minimum number of organisms is present, before reacting. These responses are diverse and include the exchange of plasmids and production of antibiotics and other biologically active molecules. Plants are susceptible to bacterial pathogens such as *Erwinia carotovora*, which produces enzymes capable of degrading its cell walls. The synthesis of these enzymes is under the control of AHLs and so they are made only once the appropriate threshold level of this chemical has been reached. The rationale behind using AHLs for plant protection is to make transgenic plants, tobacco in this case, which express this signal themselves. The consequent high level of AHL presented to the pathogenic bacteria, wrongly indicates a very high number of similar organisms in the vicinity, and triggers the bacteria into responding. As a consequence, they produce enzymes able to degrade the plant cell walls and continue infection. The plant will mount its normal response to invasion but on a far greater scale than necessary to destroy the few bacteria actually causing the infection, thus improving the plant's resistance to the disease. It seems complicated, but early research into the validity of the hypothesis (Fray *et al.*, 1999) has ultimately led to the development and successful patenting of a means to control bacterial infection (Zhang *et al.*, 2007).

Improved tolerance

Plant–microbe interactions are addressed in Chapter 10. Among the examples given are that of *Pseudomonas syringae* which colonises the surface of leaves. This example is of bacterial rather than plant modification but impinges on interaction between the two. *Pseudomonas syringae* produces a protein which promotes the formation of ice crystals just below $0\,^{\circ}$C thus increasing the risk of frost damage. Lindow *et al.* (1989) have identified and isolated the gene for this protein. They transferred it to the bacterium *Escherichia coli* to simplify the genetic manipulations. This required the deletion of sufficient regions so that a truncated, and therefore non-functional, ice mediating protein was expressed. They reintroduced this mutated gene into *Pseudomonas syringae* and selected for *ice⁻* mutants which were no longer able to produce the ice nucleating protein. Many such regimes fail in practice because it is difficult to maintain a population of mutant bacteria in a community dominated by the wild type as frequently, the latter will soon out number the mutant by competition for nutrients, since it is usually better adapted to the particular environment than the mutant. However, in this case, due to massive application of *Pseudomonas syringae ice⁻* to strawberry plants, the mutants were able to compete with the wild type and protect this particularly susceptible crop against frost damage.

Salt tolerance in tomatoes has been established by introducing genes involved in Na^+/H^+ antiport, the transport of sodium and hydrogen ions in opposite directions across a membrane. The quality of the fruit was maintained by virtue of the fact that the sodium accumulation caused by the antiport occurred in leaves only and not in the fruit (Zhang and Blumwald, 2001).

Improved tolerance to drought, salt and freezing in *Arabidopsis* has been achieved by overexpressing a protein which induces the stress response genes. However, if too much of this factor is produced, which was the case when the 35S CaMV was employed, severe growth retardation was observed. No such problem existed when instead the overexpression was under the control of a promoter which was only switched on when stressful conditions existed (Kasuga *et al.*, 1999).

Improved plants for phytoremediation

Chapter 7 mentioned the genetic modification of a poplar to enable mercury to be removed from the soil and converted to a form able to be released to the atmosphere. This process is termed 'phytovolatilisation' (Rugh *et al.*, 1998). The modification required a gene to be constructed, styled on the bacterial *mer A* gene, by making a copy reflecting the codon bias found in plants using PCR technique. The *mer A* gene is one of a cluster of genes involved in bacterial detoxification of mercury, and is the one coding for the enzyme, *mercuric ion reductase*, which converts mercury from an ionic to a volatile form. Initially the constructed *mer A* gene was expressed in *Arabidopsis thalia* (rape) where resistance to mercury was observed, and in this study, the gene was transferred by microprojectile bombardment ('gene guns') to poplar tree (*Liriodendron tulipifera*) embryogenic material. When the resulting yellow poplar plantlets were allowed to develop, they were found to exhibit tolerance to mercury and to volatalise it at 10 times the rate observed in untransformed yellow poplar plantlets. This study demonstrated the possibility that trees can be modified to become useful tools in the detoxification of soil contaminated with mercury. These studies were pursued in *Arabidopsis thalia* where it was observed that successful remediation also required the *mer B* genes coding for a *lyase* (Bizily, Rugh and Meagher, 2000).

A bacterial gene encoding pentaerythritol tetranitrate reductase, an enzyme involved in the degradation of explosives, have been transferred into tobacco plants. The transgenics have been shown to express the correct enzyme and trials undertaken to determine the extent of their ability to degrade TNT (French *et al.*, 1999).

Developments in the use of transgenic plants for bioremediation have been reviewed (Francova *et al.*, 2001; Eapen, Singh and D'Souza, 2007; Van Aken, 2009).

New products from plants

The rape plant, *Arabidopsis thalia* has become a popular choice for the production of recombinant species. One such recombinant is a rape plant, the fatty acid composition in the seed of which has been modified. It now produces triacylglycerols containing elevated levels of trierucinic acid suitable for use in the polymer industry (Brough *et al.*, 1996) and, in a separate project, polyhydroxybutyrate suitable for the production of biodegradable plastics (Hanley, Slabas and Elborough, 2000). Synthesis of the copolymer poly(3-hydroxybutyrate-*co*-3-hydroxyvalerate) by *Arabidopsis*, is another example of the application of *Agrobacterium tumefaciens* technology and the use of the 35S promoter from Cauliflower Mosaic Virus (Slater *et al.*, 1999). This copolymer can be produced by bacterial fermentation, but due to cost considerations, it is normally synthesised chemically. This example is discussed further in Chapter 10 under the umbrella of 'clean' technology. Progress in this field has been reviewed (Snell and Peoples, 2002; Mooney, 2009).

Closing Remarks

It is something of an irony that GE is virtually synonymous with biotechnology. The advances in this field have been enormous and, in many areas, are of singularly great importance, yet its impact has been much less dramatic when considering the purely environmental aspects. So much of what has been discussed in this chapter has not managed to make the wholesale transition into mainstream commercial activity and whether it ever will still remains to be seen, though the increase in patents in the field gives rise to some optimism that it might. Genomics, for example represents one fast growing set of technologies which may assume greater importance in the field in the future. It is something of a blanket term used to describe a broad swath of '-omic' disciplines, including the likes of genome architecture and sequencing, metabolomics, proteomics and transcriptomics. Of the various potentially beneficial individual -omics, perhaps toxicogenomics – the examination of toxin-induced changes in gene expression – could prove the most useful to environmental biotechnology. If it can ultimately be developed to provide effective practical biomarkers to help monitor ecological risk, it could open the way for a major decision making tool and facilitate the choice and deployment of appropriate remedial technologies. However, it remains important to remember that in any practical application of this kind, technology transfer into the essentially uncontrolled setting of a real world situation is seldom straightforward and perfecting the technique for widespread use may take a little time to achieve. Nevertheless, GE will unquestionably continue to play a definitive role in the future development of the biological sciences, though its position in respect of practical environmental biotechnology perhaps remains less clear.

References

Bajrovic, K., Erdag, B., Atalay, E.O. and Cirakoclu, B. (2001) Full resistance to Tobacco Mosaic Virus infection conferred by the transgenic expression of a recombinant antibody in tobacco. *Biotechnology and Biotechnological Equipment*, **15**, 21–27.

Beggs J.D. (1981) in *Gene Cloning in Yeast Genetic Engineering*, vol. 2 (ed. R. Williamson), Academic Press, London, New York, Toronto, Sydney, San Francisco, pp. 175–203.

Bialy, H. (1997) Comment – biotechnology, bioremediation, and blue genes. *Nature Biotechnology*, **15**, 110.

Bizily, S.P., Rugh, C.L. and Meagher, R.B. (2000) Phytodetoxification of hazardous organomercurials by genetically engineered plants. *Nature Biotechnology*, **18**, 213–217.

BMB (1995) *Biotechnology Means Business Status Report Textile and Clothing Industries*, DTI, London, p. 17.

Brough, C.L., Coventry, J.M., Christie, W.W. *et al.* (1996) Towards the genetic engineering of triacylglycerols of defined fatty acid composition: Major changes in erucic acid content at the sn-2 position affected by the introduction of a 1-acyl-sn-glycerol-3-phosphate acyltransferase from Limnanthes douglasii into oil seed rape. *Molecular Breeding*, **2**, 133–142.

Cho, H.S., Cao, J., Ren, J.P. and Earle, E.D. (2001) Control of Lepidopteran insect pests in transgenic Chinese cabbage (Brassica rapa ssp pekinensis) transformed with a synthetic *Bacillus thuringiensis* cry1C gene. *Plant Cell Reports*, **20**, 1–7.

Cook, M.A., Osborn, A.M., Bettandorff, J. and Sobecky, P.A. (2001) Endogenous isolation of replicon probes for assessing plasmid ecology of marine sediment microbial communities. *Microbiology*, **147**, 2089–2101.

Daniell, H. and Moar, W. (2007) Tobacco chloroplast transformation vectors comprising a multi-gene operon encoding a biopharmaceutical protein and a chaperonin. United States Patent 7294506.

Deák, M., Horváth, G.V., Davletova, S. *et al.* (1999) Plants ectopically expressing the iron-binding protein, ferritin, are tolerant to oxidative damage and pathogens. *Nature Biotechnology*, **17**, 192–196.

De Cosa, B., Moar, W., Lee, S.B. *et al.* (2001) Overexpression of the Bt cry2Aa2 operon in chloroplasts leads to formation of insecticidal crystals. *Nature Biotechnology*, **19**, 71–74.

Eapen, S., Singh, S. and D'Souza, S. (2007) Advances in development of transgenic plants for remediation of xenobiotic pollutants. *Biotechnology Advances*, **25** (5), 442–451.

Francova, K., Macek, T., Demnerova, K. and Mackova, M. (2001) Transgenic plants – a potential tool for decontamination of environmental pollutants. *Chemicke Listy*, **95**, 630–637.

Fray, R.G., Throup, J.P., Daykin, M. *et al.* (1999) Plants genetically modified to produce N-acylhomoserine lactones communicate with bacteria. *Nature Biotechnology*, **17**, 1017–1020.

French, C.E., Rosser, S.J., Davies, G.J. *et al.* (1999) Biodegradation of explosives by transgenic plants expressing pentaerythritol tetranitrate reductase. *Nature Biotechnology*, **17**, 491–494.

Godwin, I., Fordlloyd, B. and Newbury, H. (1992) *In vitro* approaches to extending the host-range of agrobacterium for plant transformation. *Australian Journal of Botany*, **40**, 751–763.

Grierson, D. and Schuch, W. (1993) Control of ripening. *Philosophical Transactions of the Royal Society of London*, **342**, 241–250.

Hanley, Z., Slabas, T. and Elborough, K.M. (2000) The use of plant biotechnology for the production of biodegradable plastics. *Trends in Plant Science*, **5**, 45–46.

He, M., Yang, Z.Y., Nie, Y.F. *et al.* (2001) A new type of class 1 bacterial 5-enolpyruvylshikimate-3-phosphate synthase mutants with enhanced tolerance to glyphosate. *Biochimica et Biophysica Acta-General Subjects*, **1568**, 1–6.

Hehemann, J-H., Correc, G., Barbeyron, T. *et al.* (2010) Transfer of carbohydrate-active enzymes from marine bacteria to Japanese gut microbiota. *Nature*, **464**, 908–912.

Hughes, M.A. (1996) *Plant Molecular Genetics*, Longman, pp. 77–105.

Kasuga, M., Liu, Q., Miura, S. *et al.* (1999) Improving plant drought, salt, and freezing tolerance by gene transfer of a single stress-inducible transcription factor. *Nature Biotechnology*, **17**, 287–291.

Kota, M., Daniell, H., Varma, S. *et al.* (1999) Overexpression of the *Bacillus thuringiensis* (Bt) Cry2Aa2 protein in chloroplast confers resistance to plants against susceptible and Bt-resistant insects. *Proceedings of the National Academy of Sciences of the United States of America*, **96**, 1840–1845.

Koziel, M.G., Beland, G.L., Bowman, C. *et al.* (1993) Field performance of elite transgenic maize plants expressing an insecticidal protein derived from *Bacillus thuringiensis*. *Biotechnology*, **11**, 194–200.

Le, V.Q., Belles-Isles, J., Dusabenyagasani, M. and Tremblay, F.M. (2001) An improved procedure for production of white spruce (Picea glauca) transgenic plants using Agrobacterium tumefaciens. *Journal of Experimental Botany*, **52**, 2089–2095.

Lindow, S.E., Panopoulos, N.J. and McFarland, B.L. (1989) Genetic engineering of bacteria from managed and natural habitats. *Science*, **244**, 1300–7.

Magg, T., Melchinger, A.E., Klein, D. and Bohm, M. (2001) Comparison of Bt maize hybrids with their non-transgenic counterparts and commercial varieties for resistance to European corn borerand for agronomic traits. *Plant Breeding*, **120**, 397–403.

Mooney, B.P. (2009) The second green revolution? Production of plant-based biodegradable plastics. *Biochemical Journal*, **418** (2), 219–232.

Pessi, G., Blumer, C. and Haas, D. (2002) *lac z* fusions report gene expression, don't they? *Microbiology*, **147**, 1993–1995.

Promega (1996) *Protocols and Applications Guide*, 3rd edn, Promega Corporation.

Reanney, D. (1976) Extrachromosomal elements as possible agents of adaptation and development. *Bacteriological Reviews*, **40**, 552–90.

Ruf, S., Hermann, M., Berger, I.J. *et al.* (2001) Stable genetic transformation of tomato plastids and expression of a foreign protein in fruit. *Nature Biotechnology*, **19**, 870–875.

Rugh, C.L., Senecoff, J.F., Meagher, R.B. and Merkle, S.A. (1998) Development of transgenic yellow poplar for mercury phytoremediation. *Nature Biotechnology*, **16**, 925–928.

Shah, D.M., Horsch, R.B., Klee, H.J. *et al.* (1986) Engineering herbicide tolerance in transgenic plants. *Science*, **233**, 478–481.

Slater, S., Mitsky, T.A., Houmiel, K.L. *et al.* (1999) Metabolic engineering of *Arabidopsis* and *Brassica* for poly (3-hydroxybutyrate-*co*-3-hydroxyvalerate) copolymer production. *Nature Biotechnology*, **17**, 1011–1016.

Snell, K.D. and Peoples, O.P. (2002) Polyhydroxyalkanoate polymers and their production in transgenic plants. *Metabolic Engineering*, **4**, 29–40.

Stewart, L.M.D., Hirst, M., Ferber, M.L. *et al.* (1991) Construction of an improved baculovirus insecticide containing an insect-specific toxin gene. *Nature*, **352**, 85–88.

Van Aken, B. (2009) Transgenic plants for enhanced phytoremediation of toxic explosives. *Current Opinion in Biotechnology*, **20** (2), 231–236.

Yordanova, E., Gorinova, N. and Atanassov, A. (2001) The use of cytochrome P450 monooxygenase genes to introduce herbicide tolerance in tobacco. *Biotechnology and Biotechnology Equipment*, **15**, 49–55.

Zhang, H.-X. and Blumwald, E. (2001) Transgenic salt-tolerant tomato plants accumulate salt in foliage but not in fruit. *Nature Biotechnology*, **19**, 765–768.

Zhang, L., Dong, Y., Xu, J. and Zhang, X. (2007) Control of bacterial infection by quenching quorum-sensing of plant pathogenic bacteria. United States Patent 7205452.

Zhang, L., Xu, J. and Birch, R. (1999) Engineered detoxification confers resistance against a pathogenic bacterium. *Nature Biotechnology*, **17**, 1021–1024.

Case Study 9.1   Engineered Bacterial Explosives Detection (Scotland)

According to the charity Handicap International, between 15 000 and 20 000 people are killed or injured each year by landmines and unexploded ordnance, and the majority are civilians, often trying to rebuild their lives after regional conflicts. Nearly 90 countries around the world are riddled with minefields and explosives left over from current, recent or forgotten wars, and in many cases, placement records are poor if they exist at all. In addition to the direct deaths and injuries landmines inflict, their ever-present threat to life and limb bring untold psychological, social and economic damage in their wake. Inevitably, with an estimated 110 million landmines spread across the conflict-zones of the world, the accurate and effective detection of such hidden explosives represents an essential element in any attempt at long-term reconstruction.

A team at Edinburgh University have addressed the need for safe and low-cost mine detection by developing bacteria which luminesce in the presence of explosives. Capable of being sprayed as a solution from the air, once in the soil, the microbes multiply, glowing a characteristic green colour when they come into contact with landmines, artillery shells and the like. Within just 2 hours of the ground being treated, the bacteria reveal the whereabouts of buried explosive devices, which the local population can avoid until they can be safely removed and destroyed.

The bacteria were created by the *biobricking* process, a technique increasingly used by synthetic biologists to invent new varieties of organisms expressly designed to fulfil a specific purpose, by manipulating packages of DNA. Although the underlying use of plasmids to transfer genetic material into a target organism is a long established part of traditional GE, biobricking extends its application to allow more advanced functions to be achieved – the encoding for luminescence in the presence of explosives and explosive-derived chemicals being considered a 'device' in the language of biobrick technology.

While the Edinburgh researchers have no immediate plans to make this available commercially, it undoubtedly provides a compelling first exposition of the potential of engineered micro-organisms in the role.

10

Integrated Environmental Biotechnology

The essence of environmental biotechnology as an applied science, as we have set out to demonstrate in the preceding chapters of this book, is the harnessing of pre-existing organisms and natural cycles to bring about a desired goal. Sometimes this is achieved by relatively unsophisticated means. At others it requires rather more in the way of engineering, adaptation or modification, in one form or another, to fit nature's original to the intended purpose. Thus, though the exact form of any given iteration may differ, the underlying paradigm remains the same. Applying what is effectively a naturalistic model leads to some inevitable conclusions with far-reaching implications for the future of this particular discipline.

The fundamental necessity of mutual interactions in nature is readily accepted and understood. Hence, the natural cycles obligingly dovetail together at both the gross and the microscopic levels, with interplay existing between the organism and its environment as well as between the various central metabolic pathways. Since such integration exists already between bio-processes, and these are the very stuff upon which environmental biotechnology is based, the potential for integrated applications is clear.

At its simplest, this involves the sequential use of individual technologies to provide a solution in a linked chain of successive steps, often termed a 'treatment train'. The other extreme is the wider amalgamation of larger fundamental problems and their resolutions into a single cohesive whole. This book began by looking at the key intervention areas for environmental biotechnology and defined the three legs of that particular tripod as pollution, waste and manufacturing. This theme has been further developed, to examine how old pollution can be cleaned up and how the rational treatment of solid wastes and effluent can contribute to the reduction of new pollution. So-called 'clean' technologies represent the logical end-point of this discussion, when the production processes themselves assist in the reduction of waste and the minimisation of pollution, in the ultimate integrated system.

All industrialised countries face the same three problems in attempting to marry economic growth with environmental responsibility, namely the need to marshal

Environmental Biotechnology: Theory and Application, Second Edition Gareth M. Evans and Judith C. Furlong
© 2011 John Wiley & Sons, Ltd.

material resources, deal rationally with their waste and the requirement for adequate and affordable energy. This dichotomy of desire between compromising neither commercial success nor environmental stewardship is particularly important for the long-term future of the economy. Over the years, a certain brand of extremist environmentalist thought has sought to demonise industry and commerce, decrying them and casting them in the role of enemy. This is scarcely helpful, for two reasons. Firstly, if any particular industry is actively damaging the environment, it is hardly likely to react constructively to criticism from its avowed detractors. Secondly, and perhaps much more importantly, industry in its widest sense is what has defined humanity from the outset. It accounts for what our Neolithic ancestors did, trading skins and flint axes across Europe; it is absurd to suggest that our collective future will be different. The way ahead, then, is to accept this and chart a course which, if it cannot do the most good in absolute terms, must settle for doing the least harm. In much the same way as some have vilified industry, there are those who have held the idea of a self-sustaining civilisation up to ridicule, arguing that ultimately this would have us living in mud huts, devoid of all the benefits of science and technology. The one view is as facile as the other.

The issue of sustainability has gained ever greater significance over recent years, and this seems set to continue in the future. In 1987, under the aegis of the World Commission on Environment and Development, the Bruntland Commission coined a definition of sustainable development. Their concept of an approach which 'meets the needs of the present without compromising the ability of future generations to meet their own needs' has received widespread international acceptance. The main aims have been further developed into social progress to address the requirements of all, effective environmental stewardship, the maintenance of high and stable economic growth and levels of employment and the utilisation of natural resources in a prudent fashion (DETR, 1999). These goals also tend to offer strong commercial benefits and as a result, businesses have not been slow to see their potential. In a survey undertaken by the management consultancy, Arthur D. Little, of some 500 environmental, health and safety and other business executives in North American and Europe, 95% believed sustainable development was 'important'. Around 80% said it had significant real business value, while 70% of the Europeans and more than 55% in the US reported an active sustainable development approach to strategy and operations within their organisations, for reasons of perceived business advantage. In this context, increased efficiency, competitive streamlining, better public relations, work-force awareness and rising customer expectations were all cited, while the impact of technological innovation was universally recognised.

In many respects, the move towards integration is inevitable. We cannot unscrew one leg of our tripod without unbalancing the whole structure. Sustainable development inherently demands a cogent view of resource management, and this implicitly covers materials, waste and energy. It becomes impossible to consider them in isolation. If waste becomes viewed as raw-material-in-waiting, one bridge is clear. Between waste and energy, however, the current link is incineration and, although this route will always be relevant for some unwanted materials, the situation is less than ideal. For one thing, burning denies the bridge discussed above, by allowing little or no opportunity for reclamation. If we extend this to larger environmental issues, like reducing CO_2 production and the usage of fossil fuels, biomass, and hence environmental biotechnology, comes to occupy a pivotal position in the sustainability debate.

Bioenergy

The concept of obtaining energy from biomass material was mentioned earlier, in respect of the biological waste treatment methods involving anaerobic digestion and fermentation, and represents nothing particularly novel in itself. Methane and ethanol have been long established as fuels in many parts of the world, their production and utilisation being well documented. Both of these may be described as derived fuels, biochemically obtained from the original biomass. However, to many people around the globe, the most familiar forms of biofuel are far more directly utilised, commonly via direct combustion and, increasingly, pyrolysis. Around half the world's population relies on wood or some other form of biomass to meet daily domestic needs, chiefly cooking. Estimates put the average daily consumption of such fuels at between 0.5 and 1.0 kg per person (Twidell and Weir, 1994a). This equates to around 150 W which is an apparently high figure, but one largely explained by the typical 5% thermal efficiency of the open fire method most commonly encountered.

The energy of all biofuels derives ultimately from the sun, when incident solar radiation is captured during photosynthesis, as discussed in Chapter 2. This process collects around 2×10^{21} J of energy, or 7×10^{13} W, each year, throughout the biosphere as a whole. During biomass combustion, as well as in various metabolic processes described elsewhere, organic carbon reacts with oxygen, releasing the energy once more, principally as heat. The residual matter itself feeds back into natural cycles for reuse. It has been calculated that a yearly total of some 2.5×10^{11} tonnes of dry matter circulates around the biosphere, in one form or another, of which around 1×10^{11} tonnes are carbon (Twidell and Weir, 1994b).

This relationship of energy and matter within the biospheric system, shown schematically in Figure 10.1, is of fundamental importance to understanding the whole question of biomass and biofuels. Before moving on to examine how integrated technologies themselves combine, it is worth remembering that the crux of this particular debate ultimately centres on issues of greenhouse gases and global warming. Increasingly the view of biomass as little more than a useful long-term carbon-sink has been superseded by an understanding of the tremendous potential resource it represents as a renewable energy. Able to substitute for fossil fuels, bio-energy simply releases the carbon it took up during its own growth. Thus, only 'modern' carbon is returned, avoiding any unwanted additional atmospheric contributions of ancient carbon dioxide. However, the way in which some of

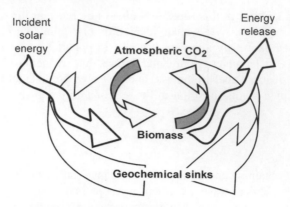

Figure 10.1   *The biomass and bioenergy cycle*

the incarnations of bio-energy are viewed in the wider context of sustainability has also gradually changed and policies and attitudes along with it. As the recent inaugural report from the United Nations Environment Programme's International Panel for Sustainable Resource Management pointed out, developing bio-energy in a truly environmentally friendly way will call for a particularly sophisticated approach (UNEP, 2009).

Derived Biofuels

Methane biogas

Biogas is a methane-rich gas resulting from the activities of anaerobic bacteria, responsible for the breakdown of complex organic molecules. It is combustible, with an energy value typically in the range of $21-28\,\text{MJ/m}^3$. The general processes of anaerobic digestion and the biochemistry of methanogenesis have been discussed in earlier sections of this book, so they will not be restated here. As mentioned previously, the main route for methane production involves acetic acid/acetate and accounts for around 75% of gas produced. The remainder is made up via methanol or carbon dioxide and hydrogen, as shown in Figure 10.2.

At various times a number of models have been put forward to aid the prediction of biogas production, ranging from the simplistic to the sophisticated. Many of these have been based more on landfill gas (LFG) generation than truly representative anaerobic bioreactors, which does lead to some confusion at times. However, it is generally accepted that the linked, interdependent curves for cellulose decomposition and gas evolution can be broadly characterised as having five principal stages, outlined below.

- Stage I: Peak biowaste cellulose loadings; dissolved oxygen levels fall to zero; nitrogen, and carbon dioxide tend to atmospheric levels.

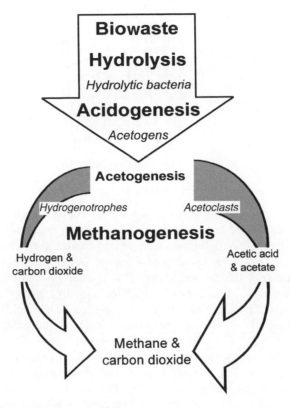

Figure 10.2   *The methanisation of biowaste*

- Stage II: Carbon dioxide, hydrogen and free fatty acids levels peak; nitrogen levels fall to around 10%; cellulose begins to be broken down.
- Stage III: Carbon dioxide decreases and plateaus to hold at around 40%; methane production commences and achieves steady state at around 60%; free fatty acids decrease to minimum levels; cellulose breakdown continues at a linear rate with respect to time; nitrogen levels fall to near zero.
- Stage IV: Carbon dioxide and methane continue in steady state at circa 40 and 60% respectively; cellulose component reduces steadily.
- Stage V: Cellulose becomes fully decomposed, ultimately leading to zero methane and carbon dioxide production; oxygen and nitrogen revert to atmospheric levels.

Although beyond the scope of the present discussion and therefore not appropriate to address fully, the position of hydrogen as a regulator of methane production warrants a brief mention. In the earlier examination of anaerobic digestion the obligate syntrophic relationship between the hydrogen producing acetogenic bacteria and the hydrogen-utilising methanogens, was described. Essentially,

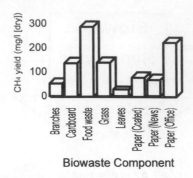

Figure 10.3   *Methane generated from biowaste components*

higher fatty acids and alcohols are converted to acetate, which requires an active population of hydrogenotrophic methanogens to ensure a low hydrogen partial pressure, avoiding the preferential production of butyric, lactic, proprionic and other acids instead of the desired acetic. This has the potential to cause higher volatile fatty acids to accumulate beyond the system's ability to self-buffer, leading to a lowering of the pH. In turn, as the increased acidity inhibits the methanogens themselves, methane production ceases and ultimately the process will collapse.

A number of different applications have developed the idea of anaerobic digestion for methane production, notably in the waste management, sewage treatment, agricultural and food processing industries. The process has also been successfully used on a relatively small scale, commonly with animal manures as its feedstock. Figure 10.3 shows an illustrative chart of methane generation for many of the common biodegradable components of MSW.

Methane has an explosive range of 5–15% by volume and a density at $20\,°C$ of $0.72\,kg/m^3$; for hydrogen the same properties lie between 4 and 74% and $0.09\,kg/m^3$ at $20\,°C$, respectively. At $20\,°C$, carbon dioxide has a density of $1.97\,kg/m^3$. The calorific value of typical biogas, consisting of about 60% CH_4, 40% CO_2, lies between $5.5–6.5\,kWh/m^3$ and it is this which makes its production attractive as a means of generating renewable energy. As was mentioned in the earlier section on anaerobic digestion, with a theoretical yield of $400\,m^3$ of biogas per wet cellulosic tonne, the prospect of high energy returns simultaneous with waste treatment has clear appeal. However, as was also pointed out in the same earlier segment, it is not feasible to optimise conditions such that high levels of both waste reduction and gas generation are deliverable. More commonly, in practice only around a quarter of the potential biogas yield is actually achieved.

Using biogas

Although biogas from engineered AD processes share many similarities with LFG, it is important to remember that it is of quite distinct quality, being much cleaner and far less contaminated by traces of other gases. LFG may contain a bewildering array of 'others', dependent on the exact nature of the waste undergoing decomposition. The list includes the likes of 1,2 dichloroethene, alkyl-benzene, butylcyclohexane, carbon disulphide, propylcyclohexan, methanethiol, decane, dichlorobenzene, undecane, ethylbenzene, dodecane, trimethylbenzene, tridecane, toluene, dimethyl disulphide, nonane and sulphur dioxide. Biogas, by contrast, is relatively pure by comparison, since the bulk of the inorganic matter and many potential pollutants are excluded from the bioreactor, either by source or mechanical separation, as part of the waste preparation process. This obviates the need for high temperature flaring, commonly used for LFG to destroy residual pollutant gases, since they are highly hostile to the fabric of any generation equipment intended to be used.

The main cause for concern in this respect is hydrogen sulphide (H_2S), which is a metabolic by-product of sulphur-reducing bacteria. Unsurprisingly, the amount present in the final gas produced depends largely on the relative abundance of sulphur containing compounds in the original biowaste. H_2S is acidic and this poses a major corrosion risk to gas handling and electrical generation equipment. Scrubbing hydrogen sulphide out of biogas is possible, but in practice it is more common to use a high-alkalinity lubricant oil which is changed often.

Biogas utilisation involves burning it, with some of the energy being transformed to electrical. There are three basic types of engine which are suitable generating motors for biogas uses, namely turbine, dual fuel and spark ignition. For each there are numbers of different manufacturers worldwide. While, clearly, it lies far outside of the scope of this book to discuss them, it is worth noting that for any given application, the type of engine used will normally be decided by a number of contextual issues. Hence, the quantity of the biogas produced, its purity, the intended life of the plant, relevant pollution controls and other similar site specific considerations will need to be considered.

Generation processes are generally relatively inefficient thermodynamically, and often much of the available energy is effectively lost as heat. However, the nature of engineered AD processes is such that there is a ready built-in demand for thermal energy to elevate and maintain the digester temperature. This may account for between 20 and 50% of the total energy produced, dependent on system specifics, in a typical temperate facility, with the remainder being available for other uses. A representative energy flow for gas engine generators is shown in Figure 10.4.

Ethanol fermentation

Fermentative processes have been described earlier, both in the general wider metabolic context and specifically in regard to their potential use in the treatment of biowaste. Fermentation produces a solution of ethanol in water, which can be further treated to produce fuel-grade ethanol by subsequent simple distillation, to 95% ethanol, or to the anhydrous form by azeotropic co-distillation using a solvent.

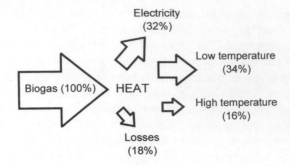

Figure 10.4   *Energy from biogas utilisation*

The relative ease with which liquid fuels can be transported and handled, coupled with their straightforward delivery to, and inherent controllability of combustion in, engines makes them of considerable importance. Ethanol is a prime example in this respect, since it can be used either as a direct replacement for petrol, or as a co-constituent in a mix. Though at $24\,GJ/m^3$, it has a lower calorific value than petrol ($39\,GJ/m^3$), in practice any performance discrepancy is largely offset by its better combustion properties.

There are thriving ethanol industries in many countries of the world, generally using specifically energy-farmed biomass in the form of primary crop plants, like corn in the US and sugarcane in Brazil. In another example of the importance of local conditions, the production costs of ethanol and the market price realised by the final fuel depend on many factors external to the technology itself. Hence the indigenous economy, employment and transport costs, government policy, taxation instruments and fiscal incentives all contribute to the overall commercial viability of the operation.

Brazil, where ethanol/petrol mixing has been routine since the 1970s is an excellent example. Although the country's use of ethanol partial substitution has a relatively long history, dating back to the 1930s, the real upsurge of acceptance of 'gasohol' lay in an unusual combination of events, partly driven by the energy-crisis of the mid-1970s. Rising oil prices, which increased by over 25% in less than two years, came at the same time as a fall in sugar revenue following a slump in the world market. The Brazilian sugarcane industry, which had shortly before invested heavily in an extensive national programme of modernisation, faced collapse. Against this background, the production of fuel from the newly available biomass crop became a sound commercial move, simultaneously reducing the country's outlay on purchased energy and buoying up one of its major industries.

The keynote of this chapter is the potential for integrating biotechnologies. In the preceding discussion of biogas, this involved the marrying together of the goals of biowaste treatment and energy production. In a similar vein, as was described in an earlier chapter, there have been various attempts, over the years, to produce ethanol from various forms of waste biomass, using naturally occurring microbes, isolated enzymes and genetically modified organisms (GMOs). The appeal to obtaining renewable energy from such a cheap and readily available source, is obvious.

In many respects, the situation which exists today with biowaste is very similar to that which surrounded Brazil's sugarcane, principally in that there is an abundant supply of suitable material available. The earlier technological barriers to the fermentation of cellulose seem to have been successfully overcome. The future of ethanol-from-biowaste as an established widespread bio-industrial process will be decided, inevitably, on the long-term outcome of the first few commercial projects. It remains fairly likely, however, that the fledgling industry will depend, at least initially, on a sympathetic political agenda and a supportive financial context to succeed. While this application potentially provides a major contribution to addressing two of the largest environmental issues of our time, energy and waste, it is not the only avenue for integrated biotechnology in connection with ethanol production.

As has already been mentioned, specifically grown crops form the feedstock for most industrial fermentation processes. The distillation which the fermentate undergoes to derive the final fuel-grade alcohol gives rise to relatively large volumes of potentially polluting by-products in the form of 'stillage'. Typically high in BOD and COD, between 6 and 16 litres are produced for every litre of ethanol distilled out. A variety of end-use options have been examined, with varying degrees of success, but dealing with stillage has generally proved expensive. Recently, developments in anaerobic treatments have begun to offer a better approach and though the research remains at a relatively early stage, it looks as if this may ultimately result in the double benefit significantly reduced cost and additional biomass to energy utilisation. The combination of these technologies is itself an interesting prospect, but it opens the door for further possibilities in the future. Of these, perhaps the most appealing would be a treatment train approach with biowaste fermentation for ethanol distillation, biogas production from the stillage and a final aerobic stabilisation phase; an integrated process on a single site. There is, then, clear scope for the use of sequential, complimentary approaches in this manner to derive maximum energy value from waste biomass in a way which also permits nutrient and humus recovery. Thus, the simultaneous sustainable management of biologically active waste and the production of a significant energy contribution becomes a realistic possibility, without the need for mass-burn incineration. In many respects this represents the ultimate triumph of integration, not least because it works exactly as natures does, by unifying disparate loops into linked, cohesive cycles.

Clearly, both AD and ethanol fermentation represent engineered manipulations of natural processes, with the activities of the relevant microbes optimised and harnessed to achieve the desired end result. In that context, the role of biotechnology is obvious. What part it can play in the direct utilisation of biomass, which generates energy by a quite different route, is less immediately apparent. One of the best examples, however, once again relates to biological waste treatment technologies, in this instance integrated with short rotation coppicing (SRC).

Short rotation coppicing

SRC differs from simple tree husbandry, being more akin to an alternative crop grown under intensive arable production. Typically using specially bred, fast-growing varieties or hybrids, often of various *Salix* or *Populus* species, SRC involves establishing plantations which are then harvested on a sustainable, basis, to provide a long-term source of biomass material for combustion. There is often a substantial land requirement associated with SRC and routinely a two to four year lead-in period. Once established, however, a yield of between 8 and 20 dry tonnes per ha per year can reasonably be expected, with a calorific value of around 15 000 MJ/tonne. Harvesting the crop forms a rotational cycle, as different sections of the plantation reach harvestable size, year on year. In this form of energy cropping, the trees themselves are effectively pruned, rather than felled, re-growth ensuring a continuing supply. Utilisation is by burning, usually in the form of chips or short lengths, most commonly for heating purposes in one form or another. In addition, the potential for producing electricity is becoming increasingly important.

The practicalities and limitations of generation from such a fuel source largely lie beyond the scope of the present work to examine. In general, though, ensuring continuity of supply and adequate production can be problematic. In addition, while much interest has been shown in the idea of using the biomass produced by a number of individual growers in a single generator, the logistics and transport costs are major obstacles to overcome. It is possible to characterise any given fuel in terms of its calorific value per unit mass, which is referred to as its *energy density* (ED). Clearly, high ED confers obvious advantages in terms of storage and delivery. Wood, however, is a relatively low ED fuel and hauling it to a centralised facility, thus, becomes costly, both in economic and environmental terms, especially over long distances. There is a clear advantage, then, in maximising the final yield of energy cropped trees and integrated biotechnology can assist in this regard.

The climate of the growing location, the irrigation needs of the particular trees being grown, the available nutrients in the soil and the management regime all play major deciding roles in the ultimate delivered biomass energy to land area ratio. While the climate must be simply accepted, the last three production variables can be optimised by judicious interventions.

The irrigation requirements of SRC have been the subject of much debate and consternation over the years. In this respect, some confusion has crept in between the needs of poplars and willows. While the former has a very deep tap root and in close planting can lower the water table by up to ten times its grassland level, the latter has a much shallower root system, making no greater a demand than a normal crop like winter wheat or sugar beet (MacPherson, 1995). Even so, at the equivalent of conventional arable requirements there still remains a large irrigation need and it is obvious that for locations with soils of poor water holding capacity, this could form a major constraint, however well suited they might otherwise be for biomass production.

Integrating biowaste products

The potential for nutrient and humus recycling from biowaste back into the soil, via composted, digested or otherwise biologically treated material was mentioned in Chapter 8. Without digressing into detailed examination of the general options open for the utilisation of such soil amendments, they do have water holding applications and form another example of the natural potential for environmental biotechnologies to self-integrate.

Much of the evidence for this has come from the field, with research conducted throughout the UK highlighting the major water holding benefits to be gained by large scale use of biowaste compost. It has been shown that at an application rate of around 250 tonnes of composted material per hectare, the land is able to hold between 1000 and 2500 tonnes of rainwater (Butterworth, 1999). Perhaps the most significant evidence in this respect comes from the trials of large-scale compost treatment in the loose, sandy soils of East Anglia, which seem to suggest that this would allow SRC crops to be grown without any further watering in all but the most exceptional of years (Butterworth, 1999). According to the same study, even under such circumstances, the additional irrigation required would be very greatly reduced. The same work established that relatively immature composts are particularly effective in this respect, as they can absorb and retain between 2 and 10 times their own weight of water. The situation appears similar for dewatered AD digestate, when applied to soil and permitted to mature *in situ*. Digestate sludges are often aerobically stabilised in a process sometimes rather inaccurately termed 'secondary composting'; this approach simply extends the same idea. The end result of this process is a high humus material, with good micro-biological activity and excellent water retaining properties, which appears to match the performance of 'true' composts at similar application levels. Moreover, it would also seem that biologically derived soil amendment materials like these, applied appropriately to soils either as a surface mulch or ploughed-in, cannot only lower supplementary watering demands enormously but also largely offset any tendency to drought-stress in the growing biomass. In addition, the leaching of nitrate from the soil is also lessened significantly.

As an aside, it is interesting to note that this ability to retain large amounts of water, together with its naturally high organic content has led to the use of compost in the construction of artificial wetlands. The US has been particularly active in this area, in part due to the fact that federal environmental regulations encourage the creation of this type of habitat as a means of water treatment. This approach, which has been discussed more fully in an earlier chapter, has as its main goal the manufacture of a wetland which behaves like a natural system in terms of both its hydrology and biology. To achieve this, a humus rich, biologically active medium, which closely replicates the normal physical and chemical properties of local soils is required. Biowaste derived composts have been found to contribute well as constituents of manufactured wetland soils, often allowing vegetation to become established on such sites more quickly than usual (Alexander, 1999).

Nutrient requirements

To return to the issue of minerals, one of the chief potential bulk end uses of biowaste-derived compost is as a horticultural amendment and fertiliser replacement. There is no clear consensus between those working in the field as to how much nutrient is removed from the system when SRC wood is harvested, estimates for nitrogen loss ranging between 30 and 150 kg/ha. A study by the UK's Forestry Commission produced figures of 135 kg/ha for nitrogen and 16 kg of phosphate, which is around one-fifth the demands made by a cereal crop. On this basis, it seems unlikely that nutrient removal would be a limit on fertile sites and certainly not for the first few harvest cycles. In the case of soils with naturally low fertility, or those which have been used for coppice cropping for some years, supplementary mineral input may well be required. Clearly, if biowaste-derived material is used for its water-holding properties, the concomitant humus and mineral donation would represent what might be described as a gratuitous benefit. Process integration in this fashion brings evident economic advantages to any commercial coppicing operation.

There is another way in which composts can help SRC. Direct competition from other plants is one of the largest factors in poor coppice crop growth and may even lead to outright failure in some cases. Uncontrolled grass or weed growth around the trees in their first season can reduce the dry matter yield by a fifth and halve their overall growth. Even after they have become properly established, weed control remains an important part of optimising the energy crop's performance, particularly where a soil's intrinsic water holding and/or nutrient levels are less than ideal. Heavy mulching has been used very successfully in many operations and, as is obvious from the previous discussions, biowaste soil amendments are ideal candidates for use in this role. It is clear that the benefits of weed suppression as a means of maximising the harvested energy yield will also apply to many other biomass crops.

Agricultural benefits of compost

In general terms, it is possible to summarise the agricultural benefits of compost as the addition of humus material and nutrients, which improve soil structure and fertility, respectively. Compost brings with it a ready made microbial community which can significantly augment the compliment already present in naturally impoverished soils. With better physical structure, aeration is improved and root growth facilitated. The ability of biowaste-derived material to contribute to a soil nutrient replacement programme, and thereby lead to a reduction in proprietary chemical fertilisers use, has been a consistent finding in numerous studies. This also represents a further prospective contribution on two relevant sides of the intervention triangle. Firstly, in reducing nitrogenous inputs, it may play a useful part in reducing the farm's pollution potential. Secondly, it becomes an example of cleaner production, since by biocycling nutrients back into the chain of biomass utility, it forms a closed loop system in respect of both minerals and energy. There may still be further 'clean' benefits to come, since research at the University of

Kassel on a range of plants, including cabbage, carrots, potatoes and tomatoes has found that the use of compost was associated with an improved nitrate to vitamin C ratio in the final product. Moreover, in structurally deficient soils especially, compost appears to produce better results than it is possible for artificial fertilisers alone to achieve. Even so, most investigations have concluded that while high application rates generally tend to give relatively big increases in crop yield, at lower levels the effect is less significant, being very largely attributable to the compost's humus enhancing effect.

Biodiesel

Returning to the central consideration of bioenergy, it would be wrong to discuss this topic without at least some passing reference to biodiesel, even though since it revolves around a chemical refining process, it is not strictly produced by biotechnology. Like the increasing number of mineral oil substitutes currently available or under development, biodiesel is derived from vegetable oils. Modern diesel engines demand a clean-burning fuel of uniform quality which can function under all expected operating conditions. One of the main advantages of biodiesel is that it can be used directly, in unmodified engines, with the additional bonus that it can perform as a single, pure fuel, or as part of a mix with its traditional counterpart, in any ratio desired. While there remains some disagreement as to the scale of the environmental benefits to be gained, especially in respect of carbon dioxide discharges, there is good evidence that particulate emissions are significantly reduced. In addition, biodiesel is claimed to have better lubricant properties and to improve the biodegradability of the conventional diesel component of a blended fuel. Various studies have concluded that biodiesel exhaust is generally less harmful to both human health and the planet. Specifically, it contains significantly lower levels of polycyclic aromatic hydrocarbons (PAHs) and nitrited polycyclic aromatic hydrocarbons (nPAHs), which is of great importance, since both groups have been identified as potential carcinogens. In laboratory tests, PAHs were reduced by between 75 and 85% (excepting benzo(a)anthracene for which the figure was around 50%) and nPAHs were also dramatically lessened. Most of the targeted nPAH compounds were present only as traces, while the highest levels reported, 2-nitrofluorene and 1-nitropyrene, were found to represent a 90% reduction over typical conventional diesel releases. Objective views of the performance of a 'new' fuel depend on such information and the National Biodiesel Board was congratulated by representatives of the House Energy and Power subcommittee for being the first industry to complete the rigorous health effects testing of the Clean Air Act.

It is not entirely without irony that in 1894, when Rudolf Diesel invented the engine which bears his name, he produced a design specifically suitable for a range of fuels, including coal dust and vegetable oil, as well as the petroleum product which is automatically associated with the device. In many respects, the current resurgence of interest in the potential of a fuel source so deeply rooted in diesel's origins might almost be described as a retrograde step in the right direction.

The great biofuel conundrum

The global demand for energy is enormous, particularly in the industrialised countries of the world and the production of biofuels has accordingly burgeoned in recent years, driven by the twin pressures of environmental and economic necessity. Ethanol production for transport tripled between 2000 and 2007, leaping from 17 billion litres to more than 52 billion; biodiesel ballooned 11-fold, to exceed 11 billion litres world-wide over the same period. Biofuels in general accounted for 1.8% of the world transport market and global investment in their production exceeded $4 billion (US) in 2007 (UNEP, 2009). According to the Worldwatch Institute, world manufacture of all forms of biofuels grew by nearly 30% in 2006 alone.

There is, however, a corollary to this meteoric rise and the fundamental question it raises is an important one. With the world's population predicted to soar to 9.4 billion by 2050, should land be given over to fuel, or food? Biofuel crops currently occupy 36 million hectares – around 2% – of the world's crop-land, but for first generation biofuels to meet 10% of the global transport demand by 2030, UN estimates suggest that perhaps as much as 500 million hectares of cropland would be required. While this could theoretically circumvent between 0.17 and 0.76 billion tonnes of fossil CO_2, the associated change in land use could predicate an overall additional $0.75-1.83$ billion tonnes of CO_2 being released (UNEP, 2009). This potential environmental own goal has not gone unnoticed, particularly in terms of fertilisers (Tilman *et al.*, 2009), indirect emissions (Melillo *et al.*, 2009) and land diversion (Searchinger *et al.*, 2008).

On a practical level, in 2007 Grain – a charity that supports the poor farmers of emergent nations – voiced concerns that the rush towards biofuels could have larger social and environmental consequences than had been thought. Shortly afterwards, the UN reported growing evidence that food prices were being driven up in poorer nations of the world as oil-rich crops were grown for fuel, while the Worldwatch Institute warned that current production methods could be exerting too heavy a price on land and water resources. It is not all doom-and-gloom for the bio-energy sector, however. Some forms of biomass derived energy, such as SRC and waste-to-biogas, for example are outside of the 'food or fuel' discussion *ab initio*, while others, such as Brazil's ethanol from the sugar cane industry are deemed to make a positive environmental contribution (UNEP, 2009). In the words of Achim Steiner, UN Under-Secretary General and Executive Director of the UN Environment Programme, 'Biofuels are neither a panacea nor a pariah but like all technologies they represent both opportunities and challenges. Therefore a more sophisticated debate is urgently needed'.

As a final and more general environmental point on this topic, as was mentioned earlier, the realisation has been growing that using biomass in a balanced way, combining its undoubted value as a carbon-sink with a progressive substitution for fossil fuels, has certain clear advantages over the sequestration-only option. Energy crop production is based on a sustainable cycle which brings benefits to the soil as well as both local biodiversity and the local economy. Land bound up in carbon sinks does not offer appreciable employment; energy farmed biomass crops can support jobs, both directly and indirectly within the region, which has evident importance to rural diversification, itself a major countryside issue. Getting the balance right between food and fuel production may be difficult, particularly in regions of the world where food, water and land resources are already under pressure, but it could clearly bring significant benefits in its wake.

Integrated Agricultural Applications

The farming industry is almost certainly about to change dramatically and the importance of novel production crops of the future will not, it seems, be limited to the energy sector. As Senator Tom Harkin of the Senate Agriculture Committee pointed out in June 2001, the potential exists for anything which can be made from a barrel of oil to be manufactured from farmed produce of one kind or another. The realisation of this is growing on a global basis and it is, therefore, highly likely that a considerable part of the forthcoming development of agricultural biotechnology will move in this direction. For reasons which should be obvious, and follow on logically from much of the preceding discussion, there is a natural fit between agricultural and environmental biotechnologies and hence, a significant potential for integration both between and within them.

Some of the ways in which this can take place in respect of biowaste-derived soil amendment products have already been described and, clearly, the advantages they convey are not limited to the particular energy crop examples cited. Before leaving this particular topic, there is another aspect of their application which is worthy of note, not least since it illustrates both integrated production and a potential means of obviating current dependence on a significant environmental pollutant.

Plant disease suppression

Intensively reared crops can suffer extensive and expensive losses resulting from plant disease infection. Until the early 1930s, crop rotation and the use of animal manures and green mulches provided the traditional protection regime; after this time, chemical fumigation became the favoured method to deal with soil-borne pathogens, which can accumulate heavily in intensive monocultures. Methyl bromide has been the main agent used, its popularity largely attributable to its ability also to destroy weeds and resident insect pests. It is, however, an indiscriminate tool, and though it has contributed directly to the commercial viability of many growers' operations, it has been implicated in ozone depletion. Accordingly, under the terms of the Montreal Protocol, it was universally phased out by 2005.

This, coupled with other fears regarding residual bromine in food and ground water, has led to bans in Germany, Switzerland and the Netherlands, the latter being, at one time, Europe's largest user of methyl bromide soil fumigation.

A number of alternative options are being explored, including soil pasturisation using steam, ultraviolet treatment and the development of resistant cultivars using both selective breeding and genetic modification. The use of compost extracts – so-called 'compost teas' – is also receiving serious consideration as a means of crop-specific disease control. Their action appears to be twofold, firstly as a protection against foliar diseases and secondly as an inoculant to restoring or enhancing sub-optimal soil microbial communities.

Research projects in Germany, Israel, Japan, the UK, the US and elsewhere have found that these extracts are very effective natural methods to suppress or control a number of plant diseases thus reducing the demand for artificial agro-chemical intervention (Table 10.1).

Direct competition with the relevant pathogen itself is one of a variety of mechanisms believed to play a part in the overall disease suppression, along with induced disease resistance, and the inhibited germination of spores. This is thought to be brought about by means of the extract's action on the surface of the leaves themselves and stimulatory effect on the associated circum-phyllospheric micro-organisms. Bacteria, yeasts and fungi present in the extracts have been shown to be active agents, while evidence points to a number of organic chemicals, including phenols and various amino acids, also having a role in the suppression effect. The exact nature of this action is not fully understood, but it would seem to be principally a biological control, since it has been known

Table 10.1   *Plant disease suppression using selected compost extracts*

Compost extract	Suppressed disease
Bark compost extract[a]	*Fusarium oxysporum* Fusarium wilt
Cattle compost extract[b,c]	*Botrytis cinerea* Grey mould of beans and strawberries
Horse compost extract[b]	*Phytophthora infestans* Potato blight
Manure-straw compost extract[d]	*Plasmopara viticola* Downy mildew of grapes *Sphaerotheca fuliginea* Powdery mildew of cucumbers *Uncinula necator* Powdery mildew of grapes
Spent mushroom compost extract[e]	*Venturia conidia* Apple scab

[a]Kai, Ueda and Sakaguchi (1990).
[b]Weltzein (1990).
[c]Elad and Shtienberg (1994).
[d]Weltzein, (1989).
[e]Cronin *et al.* (1996).

for some time that fine filtration and sterilisation by heat treatment, significantly reduce the extract's effectiveness (Trankner, 1992).

Compost teas are prepared for use by either aerated or fermented extraction methods. So-called 'fermented' extraction was the original, first developed in Germany and it is not, in fact, a fermentative process at all. Actually an infusion method, this requires a suspension of compost in water to be made, in a ratio typically around 1 : 6 by volume. The resultant mixture is allowed to stand for a given period, usually between three and seven days, then coarsely filtered prior to being used. The second method, which came out of research in Austria and the US, is more active and, with a typical cycle period of around 10–12 hours, derives the product in a much shorter time. The acceleration is achieved by increasing the oxygen transfer to the extract during formation, initially by passing water through compost, collecting the resultant liquor and recirculating it many times to concentrate and aerate. Once prepared by either method, the finished product is used as a foliar drench, typically applied to commercial crops at a rate of around 1000 l/ha (100 gal/acre).

The abilities of properly prepared biowaste composts themselves to suppress and control soil-borne plant diseases, especially where mature compost is directly mixed with the soil itself have been established (Serra-Wittling, Houot and Alabouvette, 1996). The efficacy of the protection given by this kind of incorporation with soil known to be conducive to plant pathogens has also been demonstrated. In particular, the deleterious effects of various species of Phythium, *Phytophthora* and *Fusarium* as well as *Rhizoctonia solani*, which is a major threat to many kinds of young plants, are largely suppressed or controlled. For many years the horticultural industry had been aware, anecdotally and from experience, that composted tree bark suppressed root rots caused by *Phytophthora* spp. Investigations of this revealed that plants which require the presence of vasicular arbuscular mycorrhizae (VAM), small fungi around the roots, which are intimately involved in nutrient and water uptake, grow better in blends of soil and bark compost than in methyl bromide treated soil alone. Various bacteria, fungi and yeast, largely indigenous to the original compost, have subsequently been indicated as active agents in the overall effect. The biological nature of the control has been established in laboratory trials involving heat treatment or microwave exposure of the compost. This significantly altered the microbial balance, resulting in progressive inhibition of suppression and led to 100% plant mortality in some instances. On this basis, the need to ensure that compost is not subjected to extreme temperatures during storage or transport becomes a very obvious problem to be avoided in any intended practical application.

The agricultural usage of biowaste-derived products has considerable potential, but public acceptability and quality assurance issues must be major concerns, probably more here than in any other comparable sector. In the UK alone, farmers have not been slow to learn the terrible consequences of consumer anxiety. Bovine spongiform encephalopathy (BSE) and the passions raised by genetically modified (GM) crops, or animals reared on them, entering the human food chain has focused attention ever more clearly on supply chain issues. The ramifications, in both economic and social terms, of the foot and mouth disease (FMD) outbreak, which paralysed the UK farming industry throughout 2001, has left the rural community all too well aware of the meaning of biosecurity. Since today's agri-business is so largely dominated by the demands of the supermarkets, it is neither unreasonable nor unlikely that guaranteed product quality would be a require-ment in any industry-wide standard. A clear precedent has already been set in this respect with the involvement of the British Retail Consortium in the evolution of the matrix safety code for the treatment and application of sewage sludge to agri-cultural land. The drive towards so-called 'organic' farming has already fostered a climate of proliferating, and typically widely differing, compost acceptability criteria throughout the world. This is scarcely helpful to the average would-be user of these products, often serving more to confuse than elucidate.

The prospect of widespread uptake in the growth of bio-production crops in the future has been suggested as one of the routes forward for agriculture. To echo Senator Harkin's words, farmed resources could well account for much of what is currently derived from crude oil, either directly in the chemical sense as an alternative source of the same product, or indirectly as substitutes. In the final analysis, the acceptance of the latter will, inevitably, depend on factors which are more societal and economic than scientific or technical and as such, largely beyond the scope of this book to discuss. Suffice it to say that, in any novel application, cost is a major issue and although the potential market may be enormous, the commercial benefits must be clear. As discussed before, the attitude of industry will be crucial. There is undoubtedly a strong background interest in bio-products, but machinery is often extremely expensive, and down-time is costly and inconvenient. Using a bio-engineered substitute which has not been tested and approved, often represents a huge commercial gamble, and it is a risk which few enterprises, understandably, can afford to take. It may be some time before the oft-quoted image of vast areas of transgenic crop plants growing the biological equivalents of today's petroleum based products, at no more cost than cabbages or corn, finally becomes a commercial reality. However, the beginnings are clear.

The search for a biological method of producing plastics, for example shows some promise. The ability of some bacteria to produce natural polymers has been known for some time and a number of attempts at growing plastics have been made. The products typically proved expensive, costing between three and five times as much as ordinary plastic and were generally found to be too brittle for normal use. Poly(hydroxyalkanoates) are a class of natural poly-mers with thermoplastic properties, which can be synthesised by bacteria, though the process is itself economically uncompetitive. Using green plants as plastics factories has promised much greater competitiveness, but guaranteeing the appro-priate monomer composition is not easy. One solution has been demonstrated using GM varieties of oilseed rape and cress, engineered to produce poly(3-hydroxybutyrate-co-3-hydroxyvalerate) (PHBV) a Poly(hydroxyalkanoate) with commercial applicability, within their leaves and seeds (Slater *et al.*, 1999). In effect, this is a perfect example of genetic engineering manipulating the respective strengths of contributory organisms to optimise their functional isolated potential. Inserting the bacterial genes responsible for plastic production into Arabidopsis *and* Brassica plants, avoids the expense of feed, since photosynthesis naturally provides the necessary carbon and the metabolic flow of intermediates from fatty acid and amino acid synthesis is redirected for plastic production. The PHBV biosynthesised in plant plastids appears to be of a marketable type and quality, though the yield was relatively low. The process continues to be the subject of ongoing research with the archaeon, *Haloferax mediterranei*, becoming one of the most promising candidates for commercial production of these plastics (Liu *et al.*, 2009).

Clearly, when it becomes a realistic and widespread industrial proposition, returning to the intervention triangle model to consider the agricultural benefits of specifically *environmental* biotechnology, it will represent a major advance in both 'clean' production and pollution control. While work continues on the former, in many ways the latter is already achievable.

Microbial pesticides

Chemical pesticides are problematic for many reasons. Firstly, although some degradation occurs as described in Chapter 4, pesticides are notoriously recalcitrant and consequently their use may lead to a build up of chemicals damaging to the environment. This is an increasing problem with the intensive drive towards more and more cost effective crop production. Secondly, insects are known to develop resistance to pesticides and so new and in some ways, more poisonous chemicals might be introduced to maintain the same level of effectiveness. Thirdly, chemical pesticides are rarely targeted to specific problematic species and may kill other organisms of no harm or even of some benefit to the crop plants. Balanced natural environments have an equilibrium between assailant and victim, however it may take a commercially unacceptable time for this balance to establish, sometimes incurring quite extreme swings in either direction. For example, one season may see a flourishing of citrus trees due to an outbreak of disease leading to a dearth of swallowtail butterflies, the caterpillars of which feed voraciously on citrus. The lack of insect host reduces the level of infection, leading to a recovery of the butterfly population the following year and, consequently, seriously damaged citrus trees. One means by which insect numbers are controlled in nature is by bacteria which produce toxins killing the insect which consumes it. Although this may serve to create the balance described above, it may not be sufficient to satisfy commercial crop production. Sufficient time may elapse between ingestion of the toxin by the larva and its ensuing death, for it to have caused considerable damage by feeding on the crop. Perhaps the best studied pesticidal bacterial toxin is the δ-endotoxin from *Bacillus thuringiensis*. This protein, frequently abbreviated to 'Bt toxin', is active against some members of the Lepidoptera (butterflies and moths), Diptera (flies, midges and mosquitoes) and Coleoptera (beetles) families and has been used in its native, unmodified form as a pesticide for many years. There are several strains of the bacterium each one producing a toxin active against a limited number of insect species; a relationship which is continuously evolving. Already successful and on sale to the public for domestic use, it continues to be a leading candidate for development into an even more generally useful and effective biopesticide thus reducing the dependence on chemical pesticides.

There are limitations associated with its use, all of which are being addressed in active research. These include a limited range of insects susceptible to each toxin, requiring dosing with multiple toxins, insufficient ingestion by the insect to prove lethal in a usefully short time, stability of the toxin when sprayed on crops and the development of resistance by the insect. The last stumbling block has attracted particular interest (Roush, 1994; Gould,1994; Bohorova *et al.*, 2001). The genes coding for the toxins have been isolated opening the way to their alteration and introduction into suitable 'delivery systems', either bacterial or into the plant itself, offering the plant inbuilt protection, thus attempting to overcome the various limitations introduced above.

However, even without genetic engineering, *Bacillus thuringiensis* in its native form, remains a widely used and successful product for commercial crop protection especially in 'organic' farming. In practice, the only major developments found to be necessary, are improvements to the physical formulation of the crop spray; dry, flowable formulations being an advance on the original wettable powders. Increasingly, nematodes especially *Steinernema* sp, are demonstrating great potential as biological control agents applied as sprays. It is anticipated that they will complement *Bacillus thuringiensis* in extending the spectrum of pests controlled by 'environmentally friendly' means (Knight, R, Koppert Biological Systems, personal communication).

There are other *Bacillus* species which have also been used effectively as microbial insecticides. These are *Bacillus sphaericus* which produces a toxin more potent but more specific than Bt, and *Bacillus popillae* which although not producing a toxin, kills its host by weight of bacterial numbers. The latter is active against the Japanese beetle, while the former is quite specific against mosquito larvae. Both the mosquito larvae and *Bacillus sphaericus* abound in heavily polluted water such as cesspits where the bacterium may exert control on the proliferation of mosquitoes. A different approach to microbial pesticides has been to examine the exploitation of Baculoviruses. The drive to use Baculoviruses as a means of biological pest control has dominated its research for some time but also, these viruses are recognised as vectors capable of expressing proteins of various origins at a very high level indeed and so have become enormously useful tools in the major branches of biotechnology. Several 'wild type' (not been GM) Baculoviruses are registered for use in the USA as insecticides, for example against maize bollworm (*Heliothis zea* SNPV), gypsy moth (*Lymantira dispar* MNPV), Douglas fir tussock moth (*Orgyia pseudotsugata* MNPV) and in the UK against pine sawfly (*Neodiprion sertifer* MNPV) and pine beauty moth (*Panolis flammea* MNPV). Their use as natural insecticides is now worldwide (van Beek and Davis, 2007) however, attempts to develop new, improved recombinant strains have hit licensing problems (Inceoglu, Kamita and Hammock, 2006). Members of the Baculovirus family, also called Nuclear Polyhedrosis Virus for reasons which become clear with a knowledge of their replication cycle, have been isolated from Lepidoptera, Homoptera (aphids and their relatives) and Diptera. The infectious cycle passes through a stage where several virus particles are bound together in a large crystal of protein. This protects the virus particles until the crystal is ingested by an insect where enzymes in the gut digest this polyhedrin protein releasing the viruses. These enter the insect's cells where they are uncoated, make their way into the cell nucleus, and the viral replication cycle commences. Some 12 hours after initial infection, virus particles are released which spread the infection to neighbouring cells. By 24 hours post infection, the protective polyhedrin protein coded for in the viral DNA, is being produced in sufficient quantities to start assembling the crystal structures. By this time all the insect tissues are suffering severe damage such that at the time of death, the insect is effectively a mass of virus particles surrounded by the insect's cuticle. This cadaver is eaten by birds, and consequently may be spread some distance. The virus is carried intact in the bird's gut protected in the polyhedra, which is resistant to digestion by the enzymes found in avian gut. Polyhedrin protein is made in enormous quantities in the infected cell, but since its only known function is to protect the virus *in vivo*, in the wild, then it is redundant when culturing the virus *in vitro*, in the laboratory. This being the case, the coding sequence for the polyhedrin protein may be replaced by a 'foreign' gene where under the control of the polyhedrin promoter it may, depending on the gene being replaced, have a good chance of being expressed at a very high level. This methodology used to overexpress proteins, which was described in Chapter 9, has been used with great success but increasingly in biotechnology fields other than environmental, notably pharmaceuticals.

Plant/microbe interactions

The microbiology of soils and plant microbe interactions are enormous topics worthy of the many books and research papers on the subjects, some of which are listed in the bibliography. It is not the aim of this section to give a detailed account but simply an introduction to the complexity of plant–microbe interactions in the hope of illustrating that 'no man is an island': disturbance of these interactions has its consequences. Although the term 'plant' includes all plant forms from trees to algae, the current discussion addresses interactions between higher plants and micro organisms. Such interactions fall into two basic categories: the first category being those involving microbes external to the plant, such as soil bacteria and soil fungi. The second being microbes internal to the plant, which include endophytic bacteria such as those involved in nitrogen fixation, internal fungi and plant pathogens examples of which are *Agrobacterium plasmodium* and *Agrobacterium tumefaciens* (Greene and Zambryski, 1993). The latter is now used extensively to introduce 'foreign' genes into plants as described in Chapter 9. The associations may therefore involve bacteria, fungi or viruses and in some cases, some quite complex interactions involving three or four different organisms often bringing great benefit to the plant in environments where nutrients are somewhat deficient.

Microbes external to the plant

There are clearly two distinct areas of a higher plant which are inhabited by different communities of microorganisms: above ground around and on the surface of leaves, stems, seeds and flowers and below ground in zones of increasing distance away from the root mass. These rhizospheres, or zones around the roots, which are more accurately envisaged as a continuous gradient of nutrients, are the result of plant metabolic activity constantly drawing from the surrounding soil. Nutrients may also be transferred in the reverse direction, that is to the soil, as is the case with aerating plants exemplified by *Phragmites* used in reed bed systems described in Chapter 7. It appears that colonisation of the rhizosphere by bacteria is stimulated by exudate from the plant. The first phase is attraction to the plant roots, the second is a 'settlement' phase during which bacteria grow to form colonies and the third is a 'residence' phase when a balance is established between root mass and bacterial numbers (Espinosa-Urgel, Kolter and Ramos, 2002). The microbes in the rhizosphere are, to a large extent, dependent on the plants for a supply of many useable organic substances. As plants die and decay, the components released by the degradative bacteria return to the soil and the cycle begins again. Consequently, plants affect the composition of the microbial community of the ground in which they grow, especially in soils of low fertility. Not all the compounds released from plants are stimulants of microbial growth, some plants may also produce inhibitors. The microbes themselves have an effect on the plant growth characteristics. Some release into the soil, gibberellins and cytokinins, both of which are plant growth factors, and may also affect the flow of organic compounds, termed exudate, from the plant into the rhizosphere. The rate at which exudate is transferred to the soil is affected by many parameters; the presence of surrounding soil bacteria as mentioned above, the reduction of plant mass by harvesting from above ground level, and environmental changes, for example variations in light or temperature. Both bacteria and fungi contribute to the microbial population in the rhizosphere. Associations of fungi with roots of vascular plants, called mycorrhizae, are quite common and may in some cases be very beneficial to the plant. They may be external, ectomycorrhizal, or internal, endomycorrhizal. Ectomycorrhizal associations more commonly occur in temperate regions and often in beech, oak, birch and coniferous trees. Their association involves a limited penetration of the root cortex by the fungi growing as a covering around the tip of the root. They aid the growth of the plants as a result of their mycelia reaching far out into the surrounding soil, thus assisting the plant in nutrient uptake. This quality has received commercial attention. The effect on plant growth and subsequent predation by insect larvae, of some species of fungus, for example *Pisolithus tinctorius*, has received particular attention (Rieske, 2001).

Bacteria have been found capable of encouraging this association, earning themselves the title of 'micorrhizal helper bacteria'. Clearly, anything which increases the efficiency of nutrient uptake by crop plants reduces the requirement for the addition of artificial fertilisers and thus reduces the potential for agri-chemical environmental disturbance.

The influence of microbes on the welfare of plants is not confined to the ground and may even affect the weather. An often quoted example is that drawn from *Pseudomonas syringae* which produces a protein known to act as a point of nucleation of ice crystals. Plants which harbour this bacterium run an increased risk of frost damage especially if their tissue is particularly susceptible as is the case with strawberries. *P. syringae* has been subjected to genetic engineering which successfully reduced the problem. A description of the project is given in Chapter 9.

Plant–microbe interactions are becoming recognised as having an immediate and direct importance to human health in the role they can play in reducing the effect of 'sick buildings'. They occur principally because these buildings are closed systems in which people work, breathing in volatile components from plastics, paint, chemicals used in office machinery such as photocopiers and printers and a range of other sources including furnishing manufactured using synthetic materials. Bacteria resident in the soil of potted plants in the office are able to degrade many of these volatiles, which include phenolics, formaldehyde and trichloroethylene, thus improving the air quality. The plants themselves contribute to this improvement, not only by supporting the rhizo-sphere microbial community, but also by producing oxygen during photosynthesis described in Chapter 2. In essence, this is root zone phytotechnology on a small scale.

Although straying away slightly from the remit of this section, it is interesting to point out that soil microbial activity has a major influence in the balance of stable atmospheric gases. These include the greenhouse gases, carbon dioxide, nitrous and nitric oxide and methane, so called because they trap heat re-emitted by Earth from energy radiated by the sun. The atmospheric balance of less stable gases which include ammonia, hydrogen sulphide and dimethylsulphide are also subject to microbial activity, as will be apparent from an understanding of the foregoing chapters on metabolism. A final word on soil microbes concerns the degradation of lignin. This is a major constituent of woody plant material and is recalcitrant to degradation. However, filamentous fungi are responsible for its degradation worldwide, augmented in the tropics by bacteria living in the gut of termites. This degradation requires the presence of oxygen, hence wood residing in anaerobic conditions is somewhat protected. Clearly, should the water table drop exposing to air, for example the wood pilings supporting buildings, there is a danger of invasion by filamentous fungi able to degrade lignin and thus weaken the building structure. This also explains in part the necessity to aerate a compost heap containing any woody plant material to allow invasion by filamentous fungi capable of degrading lignin.

Microbes internal to the plant

Two categories fulfil this description. The first are the internal fungi or endomy-corrhizae, referred to in the earlier section, together with the endophytic bacteria and the second comprises plant pathogens, which may be bacterial or viral in form. Although the term 'endophytic' seems unambiguous, it is sometimes used to describe only bacteria which may be isolated from plants which have been superficially cleansed with disinfectant, or isolated from within plant tissue and which cause no discernible harm to the plant. Thus defined, plant pathogens are excluded from this description. There are endophytic bacteria, called commensals, which neither benefit nor harm the plant, but there are also those which are beneficial to plant growth. These are symbiots which achieve this status either by promoting plant growth or by protection against plant pathogens.

Symbiotic nitrogen fixation

The classic example of plant growth stimulation by plant/microbe symbiosis is nitrogen fixation by *Rhizobium* bacteria within plant root tissue. A full exposition of nitrogen fixation may be found in many classical textbooks and so will be described here only in outline. The process of root nodule formation is shown diagrammatically in Figure 10.5. In total, there are a limited number of organisms able to fix nitrogen, all of which are prokaryotes. All living organisms are dependent ultimately on such organisms, due to a universal and essential requirement for nitrogen, normally in the form of nitrate, ammonia or ammonium ion or as amino acids from which the amino group may be transferred as required. Nitrogen is fixed by reduction to ammonia either by free living organisms or by plant symbiots. In both cases, it is essential to have an oxygen free environment as the enzymes involved in the process are irreversibly inactivated by the presence of oxygen. Of the free living organisms able to fix nitrogen, some achieve this naturally while others have to create such an environment. *Clostridium* and *Klebsiella* achieve this since they are both anaerobes and so are already adapted to life in an oxygen free atmosphere, while *Cyanobacteria* and *Azobacter* have developed means of creating one for themselves. *Azobacter* does this by having a very high oxygen consumption rate thus effectively creating an oxygen free environment. Other nitrogen fixing bacteria include the filamentous bacteria such as the *Corynebacterium* species, and photosynthetic bacteria referred to elsewhere, such as *Rhodospirillum*. The latter makes use of photosynthesis to provide the energy for these reactions and so is presented with the problem of removing the oxygen produced during photosynthesis away from nitrogen fixation reactions. Although these free living organisms have a vital role to play in their particular niches, approximately 10 times more nitrogen is fixed by plant symbiots. Presumably this is because the plant is better able to provide the necessary levels of ATP to meet the high energy demands of the process than are free living bacteria. In addition, the plant supplies the endophytes with dicarboxylic acids, such as malate and succinate and other nutrients, like iron, sulphur and molybdenum which is a component of the nitrogen fixing enzymes. It also provides its nitrogen-fixing symbiots with an oxygen free environment, as described later in this discussion. The analogous chemical process has an enormous energy requirement to achieve the necessary high temperature, in the region of 500 °C, and a pressure in excess of 200 atm. This is part of the reason why manufacture of fertilisers for agricultural purposes, which is in effect the industrial equivalent of the biological process, is a drain on natural resources. Together with the unwelcome leaching of surplus fertiliser into waterways, causing algal blooms among other disturbances, this makes the widespread application of fertiliser recognised as a potential source of environmental damage. It is understandable to see a drive towards the engineering of plants both to increase the efficiency of nitrogen fixation, which is estimated at being 80% efficient, and to extend the range of varieties, and especially crop species, which have this capability. It is also worth noting that unlike superficially applied fertilisers which may exceed locally the nitrogen requirement and so leach into the surrounding waterways, nitrogen fixation by bacteria occurs only in response to local need and so is very unlikely to be a source of pollution.

The following brief description of the required interactions between plants and microbes, the example used here being *Rhizobium*, should serve to illustrate why such a goal is very difficult to achieve. Firstly, the plant is invaded by a member of the *Rhizobium* family of free living soil bacteria. There is a specific relationship between plant and bacterium such that only plants susceptible to that particular member of the group may be infected. Genes involved in the infection process and nodule formation called *nod* genes are coded for by the bacterium. The *nod* genes are activated by a mixture of flavonoids released by the plant into the region around the roots and thus the plant signals to the bacterium its receptiveness to be infected.

After infection through the root hairs, the multiplying bacteria find their way into the cells of the inner root cortex. They are drawn into the cell by endocytosis shown in Figure 10.5 and so are present within the cell bounded by plant cell membrane. This structure then develops into nodules containing the nitrogen fixing bacteria. Several changes to the plant then ensue including the synthesis of proteins associated with the nodule, the most abundant of which is leghaemoglobin, which may reach levels of up to 30% of the total nodule protein. The genes coding for this protein are partly bacterial and partly plant in origin and so exemplify the close symbiosis between the two organisms. The expression of leghaemoglobin is essential for nitrogen fixation, since it is responsible for the control of oxygen levels. The enzymes for fixation are coded for by a plasmid of *Rhizobium* and are referred to as the *nif* gene cluster. There are two components each comprising a number of genes. One component is *nitrogenase reductase* the function of which is to assimilate the reducing power used by the second component, *nitrogenase*, to reduce nitrogen to ammonium ion. Expression of the *nif* genes is highly regulated in all organisms studied to date. In addition to the *nif* and the *nod* genes, *Rhizobium* also carries additional genes which are involved in the fixation of nitrogen called the *fix* genes. Once

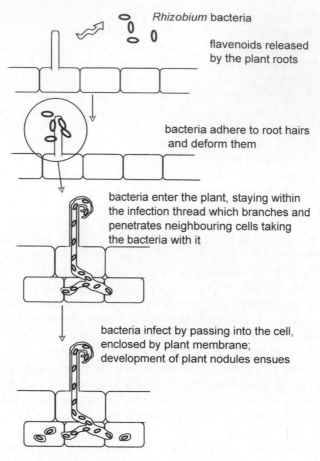

Figure 10.5   *Root nodule formation*

ammonia, or the ammonium ion, has been formed it may be transferred to the amino acid, glutamate, to produce glutamine. This transfer is frequently invoked throughout metabolism. This is not the only route for the assimilation of the amino group into metabolic pathways; the synthesis of the purine derivative, allantoin or allantoic acid, being a less heavily used alternative. These pathways are described in Chapter 2. Nitrogen fixation involving *Rhizobium* has been the focus of the preceding discussion, but the family of aerobic, Gram positive bacteria, actinomycetes, is another group carrying out the same function using a similar mechanism. These may form a network of aerial or substrate hyphae resembling a structure commonly associated with that of fungi. These bacteria may form a close association, called actinorhizae, with the roots of a number of plants which tend to be woody, or shrub like in nature.

Endomycorrhizae

It is not only bacteria which may live within plants, but some types of specialised fungi may also occupy such a niche. Their presence is fairly widespread and may be found in various grasses and a wide range of commercial crop plants including tomatoes, apples, beans, wheat and corn. One type of plant-fungus association, actinorhizae, has already been mentioned in Section 10.4.4. The fungal hyphae penetrate the plant cells where a variety of structures may develop such as swellings or the development of coils or small branches. Vesicles and arbuscules which are branched structures reminiscent of a tree, are common features of this invasion. Despite how this description may appear, such association of plant with fungus may be very beneficial to the plant. In exchange for energy derived from the plant through photosynthesis, the fungus may enhance the supply of available nutrients to the plant under conditions of relatively high humidity and conversely, in dry conditions, the fungus may help the plant in the uptake of water. In addition, some fungi have been found to afford protection to the plant in cases where the fungi produce alkaloids rendering the plant less susceptible to attack by chewing insects.

Plant pathogens

There are many bacteria, fungi and viruses which may infect a plant and cause disease, fungi being one of the major causes of plant disease. Most microorganisms will not be pathogens for a particular plant as the manifestation of disease requires some quite specific reactions and responses between host and infectious agent. Infection elicits numerous responses some of which may be quite complicated, and so has been the centre of some fairly intensive research. The result of this has been the identification and isolation of plant genes involved in resistance to pathogens and pathogen virulence genes. Manipulation of these to reduce the level of environmentally damaging chemicals to protect crop plants is an active area of research and the manipulation of these genes is discussed in the previous chapter. This section is restricted to two examples, one bacterium and one virus, which are chosen because of their relevance to plant genetic engineering described in Chapter 9. The bacterium is *Agrobacterium tumefaciens* and the virus is Cauliflower Mosaic Virus (CaMV).

Agrobacterium tumefaciens

A feature of infection caused by this Gram positive organism, is a tumour-like growth, seen as a crown gall in plants. This is the consequence of injection by the bacterium of a small piece of DNA which carries on it genes which code for opines which encourage further invasion of the plant by the bacterium, and for plant growth hormones including auxin whose activity stimulate plant growth thus produces the characteristic tumour. It is interesting to note this example of gene transfer from prokaryote to eukaryote occurring with moderate frequency in nature. The genes coding the information required for the insertion of this small piece of DNA including the insertion sequence itself, are carried on a plasmid, described in Chapter 2, called Ti plasmid whose structure is shown in Figure 9.4. The process of infection is stimulated by exudate from a plant which has been wounded by some means not necessarily the result of infection, for example by frost damage. The *vir* genes are activated leading to nicks being introduced at the borders of the T-DNA leading to the release of one of the strands of the double stranded DNA. A copy is made of the remaining strand by the usual methods of DNA synthesis and repair thus restoring the *status quo* of the Ti plasmid. The single stranded piece of T-DNA is free to be transferred into the plant cell through a wound site and on into the plant cell nucleus. Here the complementary strand is synthesised in the normal way using the plant cell enzymes and the resulting double stranded DNA integrates into the plant genome. The T-DNA comprises three genes including those for plant hormones and opines, as mentioned above. Once the T-DNA is integrated into the host plant cell genome, these genes may be expressed leading to the establishment of crown gall disease. *Agrobacterium tumefaciens* then contains a very effective delivery system for bacterial genes into a plant cell; a natural process which is utilised by genetic engineers to introduce 'foreign' genes into plants as described in Chapter 9.

Cauliflower mosaic virus

The study of plant viruses has, historically, lagged behind that of animal or bacterial viruses due to difficulties in their culture and isolation. While some may be grown in isolated plant cultures, many require the whole plant and some also require an insect intermediate either simply as a means of transmission between plants, or additionally as a site for virus replication. Transfer of viruses between plants by insects is the most common means, especially by insects which penetrate and suck plants, aphids being an example. However, there are other routes such as transfer by plant parts through infected seed, tubers or pollen. Other agents are those able to penetrate plant tissue which include soil nematodes and parasitic fungi. The genetic material of most plant viruses is RNA, either double or single stranded. Caulimovirus, or more commonly, CaMV is unusual in having a DNA genome. This has proven very fortuitous for genetic engineering as it possesses two very strong constitutive promoters, the 35S and the 19S; for an explanation of the term 'promoter' see Chapter 2. Since these promoters originated from a plant virus, any construct made with the intention of expressing the 'foreign' gene in a plant has a higher likelihood of the signals being recognised by the transcription machinery of the plant than if they were derived from, for example bacterial promoters. These promoters have proved very successful and particularly the 35S, often designated in publications as '35S CaMV', has become almost the archetype promoter to drive the expression of 'foreign' genes in plants. Examples of these are given in Chapter 9. However, while genetic engineering, building on these and other natural abilities, may permit novel technological approaches to emerge, which have major potential relevance to the environmental sphere, they can be of somewhat limited commercial application. At present, there seems to be much more scope, at least in practical terms, for the bundling of existing technologies into treatment trains, or the re-entry of post-processed biological material into the chain of commercial utility.

Closing Remarks

Biotechnological integration often permits a number of key environmental concerns to benefit and as we begin the second decade of the third millennium, there is little doubting that we face some serious environmental concerns indeed. As the dust begins to settle from COP-15 and the controversies of Climate-Gate, it seems that issues far more important than whether a view is 'orthodox' or 'heretic' are emerging that usefully begin to move the debate beyond puerile accusations of denial or belief. The future does not lie in discussions as to how anthropogenic climate change truly is, or is not; it lies in a paradigm shift in our stewardship of the planet and a better realisation of the need to conserve, sustain and marshal our resources with intelligence and probably greater humility.

'The world must come together to confront climate change. There is little scientific dispute that if we do nothing, we will face more drought, famine and mass displacement that will fuel more conflict for decades. For this reason, it is not merely scientists and activists who call for swift and forceful action – it is military leaders in my country and others who understand that our common security hangs in the balance'. So said President Obama in his Nobel Prize acceptance speech, encapsulating the real crux of the huge challenge that faces us all as water, food and energy resources come under increasing pressure across the globe.

When fellow Nobel Peace Prize Laureate, Norman Ernest Borlaug, showed the world how to improve agricultural yield back in 1970, the population that needed to be fed stood at only 3.7 billion; today it is creeping inexorably towards 7 billion and will be fast approaching 9.5 billion by the middle of the century. Currently, nearly 50% of the world's food supply is grown in 20% of the world's cultivated land, that is irrigated. It is thirsty work, with irrigation in some parts of the world accounting for 80% or more of the total water used. Small wonder then, that in 2009, the International Water Management Institute warned of potential future food shortages in Asia, in the absence of significant reform of the continent's agricultural water usage. One thing is clear; Asia is certainly not alone.

Environmental biotechnology unquestionably has the potential to make a lasting contribution to meeting future demands for food, fuel and water; the technologies are already here and others are waiting in the wings. In the final analysis, life is enormously robust and resilient, not perhaps at an individual level, but certainly on a gross scale. Living things, and most especially microbes, have colonised a truly extensive range of habitats across the planet, and some of these are, as has been discussed, extremely challenging places and surprising new examples are being found all the time. Recently, for example, three new species of the phylum Loricifera were discovered in the sediments of the deep anoxic hyper-saline L'Atalante basin of the Mediterranean – the first metazoans known to science that live their entire lives in conditions of permanent anoxia (Danovaro *et al.* 2010). Such biological diversity, combined with the lengthy history of bacteria and archaea, which has equipped many species with an amazing array of residual metabolic tools, adds up to a remarkable reservoir of capabilities which may be of use to the environmental biotechnologist. Meeting the challenges to come may well all boil down to how well we can integrate our thinking as much as our technologies and capitalise on all that potential.

References

Alexander, R. (1999) Compost Markets Grow with Environmental Applications. *BioCycle Magazine* (April 1999), p. 48.

van Beek, N. and Davis, D.C. (2007) Baculovirus insecticide production in insect larvae, *Baculovirus and Insect Cell Expression Protocols*, Methods in Molecular Biology, Vol. 388, Springer, New Jersey, pp. 367–378.

Bohorova, N., Frutos, R., Royer, M. *et al.* (2001) Novel synthetic *Bacillus thuringiensis* cry 1B gene and the cry 1B-cry1 Ab translational fusion confer resistance to southwestern corn borer, sugarcane borer and fall armyworm in transgenic tropical maize. *Theoretical and Applied Genetics*, **103**, 817–826.

Butterworth, W. (1999) A top idea that holds water. *Wet News, (Water and Effluent Treatment News)*, **5** (17), 4.

Cronin, M., Yohalem, D., Harris, R. and Andrews, J. (1996) Putative mechanism and dynamics of inhibition of the apple scab pathogen *Venturia inequalis* by compost extracts. *Soil Biology and Biochemistry*, **28**, 1241–1249.

Danovaro, R., Dell'Anno, A., Pusceddu, A., Gambi, C., Heiner, I. and Kristensen, R.M. (2010) The first metazoa living in permanently anoxic condition. *BMC Biology*, **8**, 30.

DETR (Department of the Environment, Transport and the Regions) (1999) A Way with Waste: A Draft Waste Strategy for England and Wales, Part 1, p. 9 (citing the UK Government's Sustainable Development Strategy, A Better Quality of Life, launched by the Rt. Hon. John Prescott MP, Deputy Prime Minister, in May of the same year).

Elad, Y. and Shtienberg, D. (1994) Effect of compost water extracts on grey mould (*Botrytis cinerea*). *Crop Protection*, **13**, 109–114.

Espinosa-Urgel, M., Kolter, R. and Ramos, J.-L. (2002) Root colonization by *Pseudomonas putida*: love at first site. *Microbiology*, **148**, 1–3.

Gould, F. (1994) Potential and problems with high-dose strategies for pesticidal engineered crops. *Biocontrol Science and Technology*, **4**, 451–461.

Greene, E.A. and Zambryski, P.C. (1993) *Agrobacteria* mate in opine dens. *Current Biology*, **3**, 507–509.

Inceoglu, A.B., Kamita, S.G. and Hammock, B.D. (2006) Genetically modified baculoviruses: A historical overview and future outlook, in *Advances in Virus Research*, Insect Viruses: Biotechnological Applications, Vol. 68 (eds

B.C. Bonning, K. Maramorosch and A.J. Shatkin), Elsevier, Amsterdam, pp. 323–360.

Kai, H., Ueda, T. and Sakaguchi, M. (1990) Antimicrobial activity of bark-compost extracts. *Soil Biology and Biochemistry*, **22**, 983–986.

Liu, X.-W., Wang, H.-H., Chenc, J.-Y. *et al.* (2009) Biosynthesis of poly(3-hydroxybutyrate-*co*-3-hydroxyvalerate) by recombinant *Escherichia coli* harboring propionyl-CoA synthase gene (*prpE*) or propionate permease gene (*prpP*). *Biochemical Engineering Journal*, **43**, 72–77.

MacPherson, G. (1995) *Home Grown Energy from Short-Rotation Coppice*, Farming Press Books, Ipswich, p. 175.

Melillo, J.M., Reilly, J.M., Kicklighter, D.W. *et al.* (2009) Indirect emissions from biofuels: how important? *Science*, **326**, 59581397–59581399.

Rieske, L.K. (2001) Influence of symbiotic fungal colonization on oak seedling growth and suitability for insect herbivory. *Environmental Entomology*, **30**, 849–854.

Roush, R.T. (1994) Managing pasts and their resistance to *Bacillus thuringiensis* – can transgenic crops be better than sprays. *Biocontrol Science and Technology*, **4**, 501–516.

Searchinger, T., Heimlich, R., Houghton, R.A. *et al.* (2008) Use of U.S. croplands for biofuels increases greenhouse gases through emissions from land use change. *Science*, **319**, 58671238–58671240.

Serra-Wittling, C., Houot, S. and Alabouvette, C. (1996) Increased soil suppressiveness to *Fusarium* wilt of flax after addition of municipal solid waste compost. *Soil Biology and Biochemistry*, **28**, 1207–1214.

Slater, S., Mitsky, T., Houmiel, K. *et al.* (1999) Metabolic engineering of *Arabidopsis* and *Brassica* for poly(3-hydroxybutyrate-co-3-hydroxyvalerate) copolymer production. *Nature Biotechnology*, **17**, 1011–1016.

Tilman, D., Socolow, R., Foley, J.A. *et al.* (2009) Beneficial biofuels – the food, energy, and environment trilemma. *Science*, **325**, 5938270–5938271.

Trankner, A. (1992) Use of agricultural and municipal organic wastes to develop suppressiveness to plant pathogens, in *Biological Control of Plant Diseases: Progress and Challenges for the Future* (cds E. Tjamos, G. Papavizas and R. Cook), Plenum, New York, pp. 35–42.

Twidell, J. and Weir, T. (1994a) *Renewable Energy Resources*, Chapman & Hall, London, p. 291.

Twidell, J. and Weir, T. (1994b) *Renewable Energy Resources*, Chapman & Hall, London, p. 281.

UNEP (2009) United Nations Environment Programme's International Panel for Sustainable Resource Management Towards Sustainable Production and Use of Resources: Assessing Biofuels.

Weltzein, H. (1989) Some effects of composted organic materials on plant health. *Agriculture, Ecosystems and Environment*, **27**, 439–446.

Weltzein, H. (1990) The use of composted materials for leaf disease suppression in field crops. crop protection in organic and low-input agriculture. *British Crop Protection Council Monographs*, **45**, 115–120.

Case Study 10.1   Microbial Control and Cell Signalling (Australia)

With so much of environmental biotechnology depending on microbes at the functional level, the prospect of being able to adjust the metabolic rate of bacteria, to either speed things up or slow them down, presents a powerful tool. Clearly such microbial control has potential applications in a wide range of industries and activities, but arguably its most successful exposition to date has been found in cleaning up sewer systems.

Sewage itself is inevitably rich with a wide variety of bacteria and their combined microbial biomass tends to accumulate in the bends and dips of the sewerage pipes that convey effluent to wastewater treatment plants. Over time, these accretions eventually form biofilms and subsequently give rise to both foul-smelling gases and corrosion, accounting for extensive odour control expenditure and costly infrastructure maintenance.

To combat this problem, the Australian company Biosol researched the cell signalling chemicals found in nature, ultimately harnessing them to develop commercial products capable of precisely influencing bacterial activity. By effectively placing the microbes in a state of hibernation, sewage that has been dosed travels inoffensively along sewer pipes, incapable of forming the biofilms that typically prove so problematic.

However, once inside the treatment works, a quiescent bacterial population within the flow would obviously be disastrous. Reversing the process with the judicious dosing of 'speed-up' signals, microbial breakdown is accelerated and enhanced, to provide a faster processing, higher treatment throughput, a better quality final effluent and a lower residual biosolids load. It also offers the potential of an improved methane yield, which can in turn help power the plant, or alternatively be made available for export off-site.

The approach is based in 'quorum sensing' – the means by which bacteria become aware that their population level is at a particular trigger point and then naturally modify and co-ordinate their behaviour accordingly. Much of the research has centred on biomedical applications, particularly restoring antibiotic effectiveness, but Biosol's breakthrough has demonstrated the enormous benefits this technology could offer a wide range of environmental problems too.

Case Study 10.2   Integrated Sludge Digestion and Biomass Energy Recovery (Wales/the Netherlands)

Renewable energy has assumed burgeoning importance over recent years and with the water industry facing calls to simultaneously cut its carbon footprint and improve treatment levels, any technology that helps achieve both is doubly welcome.

Aiming to generate over 5 MW of 'green' power – and cutting Dŵr Cymru/Welsh Water's carbon footprint by an estimated 35 000 tonnes of CO_2 equivalent annually – two new advanced bio-digestion plants are being built in Cardiff and Afan. Gouda-based Imtech was selected as the operation's process partner, for a solution that involves thermal hydrolysis of the sludge, heating it to 170 °C for pasteurisation before it enters the anaerobic digestion phase.

Continued on page 267

___ Continued from page 266 _____

One significant advantage of this process is that the yield of methane-rich biogas is high, which enables maximum recovery of the biomass energy to be achieved. Feeding this into five state-of-the art combined heat and power cogeneration engines, rated at 1.4 MW apiece, will produce up to 5.5 MW of electricity. The combined heat and power produced will allow each site to be effectively energy self-sufficient, have the additional capacity to be able to meet the needs of the adjoining wastewater treatment and, in the case of the Afan site, also export up to 1 MW of electricity to the grid.

Aside of their role in energy production – tipped to cut Dŵr Cymru's overall carbon footprint by around 15% – the plants will also provide a beneficial use for the sewage sludge product. After dewatering, the pathogen-free sludge-cake will be distributed across South Wales for use as an agricultural fertiliser/soil enhancer, ensuring that this multifaceted project also dovetails into the ongoing regional attempts at improving sustainability.

Between cutting energy usage, supplying more consumers and improving water/wastewater treatment quality, the water industry as a whole has been asked to perform an increasingly awkward balancing act. Accordingly, initiatives like this one, in which biotechnologies for effluent improvement, beneficial waste recycling and bio-energy generation combine to provide a single comprehensive solution, may become increasingly common in future.

Bibliography and Suggested Further Reading

Allison, D.G., Gilbert, P., Lappin-Scott, H.M. and Wilson, M. editors (2000) *Community structure and co-operation in biofilms* Fifty-ninth symposium of the Society for General Microbiology held at the University of Exeter, September 2000. Cambridge University Press, Cambridge.

Barrow, G.I. and Feltham, R.K.A. (1993) *Cowan and Steel's Manual for the Identification of Medical Bacteria*, 3rd edition, Cambridge University Press, Cambridge.

Bhattacharyya, B.C. and Banerjee, R. (2007) *Environmental Biotechnology*, Oxford Higher Education, India.

Davidson, J.N. (1972) *The Biochemistry of the Nucleic Acids*, 7th edition, Chapman and Hall, London.

Dwivedi, P. (2007) *Biodiversity and Environmental Biotechnology*, Scientific Publishers Journals Dept., India.

Evans, G.M. (2001) *Biowaste and Biological Waste Treatment*, James and James, London.

Glazer, A.N. and Nikaido, H. (1995) *Microbial Biotechnology Fundamentals of Applied Microbiology*, WH Freeman and Company, New York.

Hardman, D.J., McEldowney, S. and Waite, S. (1994) *Pollution: Ecology & Biotreatment*, Longman Scientific & Technical, Harlow.

Hughes, M.A. (1996) *Plant Molecular Genetics*, Longman, Harlow.

Larson, R.A. and Weber, E.J. (1994) *Reaction Mechanisms in Environmental Organic Chemistry*, Lewis Publishers, Boca Raton.

Lehninger, A.L. (1975) *Biochemistry*, 2nd edition, Worth Publishers Inc, New York.

MacPherson, G. (1995) *Home Grown Energy from Short-Rotation Coppice*, Farming Press Books, Ipswich.

Mandelstam, J. and McQuillen, K. (1973) *Biochemistry of Bacterial Growth*, 2nd edition. Blackwell Scientific Publications, Oxford.

Nelson, D.L. and Cox, M.M. (2000) *Lehninger Principles of Biochemistry*, 3rd edition. Worth Publishers Inc, New York.

Polprasert, C. (1995) *Organic Waste Recycling*, John Wiley & Sons, Chichester.

Environmental Biotechnology: Theory and Application, Second Edition Gareth M. Evans and Judith C. Furlong
© 2011 John Wiley & Sons, Ltd.

Prescott, L.M., Harley, J.P. and Klein, D.A. (1996) *Microbiology*, 3rd edition, Wm C. Brown Publishers, Dubuque IA.

Scragg, A. (2005) *Environmental Biotechnology*, 2nd edition. Oxford University Press, Oxford.

Twidell, J. and Weir, T. (1994) *Renewable Energy Resources*, Chapman & Hall, London.

Vallero, D. (2010) *Environmental Biotechnology: A Biosystems Approach*, Academic Press, London.

White, A., Handler, P., Smith, E.L. (1968) *Principles of Biochemistry*, 4th edition, McGraw-Hill Book Company, New York.

Index

Page numbers in *italics* refer to Figures.

Environmental Biotechnology: Theory and Application, Second Edition Gareth M. Evans and Judith C. Furlong
© 2011 John Wiley & Sons, Ltd.